Community Gardening in an Unlikely City

Community Gardening in an Unlikely City

The Struggle to Grow Together in Las Vegas

Tyler Schafer

LEXINGTON BOOKS
Lanham • Boulder • New York • London

Published by Lexington Books
An imprint of The Rowman & Littlefield Publishing Group, Inc.
4501 Forbes Boulevard, Suite 200, Lanham, Maryland 20706
www.rowman.com

6 Tinworth Street, London SE11 5AL, United Kingdom

British Library Cataloguing in Publication Information Available

Library of Congress Control Number: 2020944800
ISBN 978-1-7936-2312-6 (cloth)
ISBN 978-1-7936-2313-3 (electronic)

For all the people who have contributed their blood, sweat, and tears in making Vegas Roots Community Garden the special place that it is, and to those whose paths will eventually lead them there.

Contents

List of Figures and Table ix

Acknowledgments xi

1 Sowing Seeds of Community and Sustainability in a Wasteland 1

2 "We Have a Community Garden in Las Vegas?" Struggling with Segregation and Sustainability in Sin City 23

3 Miracle in the Mojave: Spiritual Place Narratives and Cultivating Community 49

4 Talking the Talk: Performativity and the Cultural Production of a Community Garden 65

5 "We Have Everything Else, but We Have No Foundation": The Impact of Strategic Choices on Collective Identity Formation 95

6 It's for Everyone, It's *for* No One: Explicit Inclusivity, Implicit Exclusivity, and the Boundaries of Community 119

7 Committing to Community 143

Appendix—Urban *Field* Work: Laboring, Learning, and Lounging with Others in a Community Garden 159

Bibliography 173

Index 185

About the Author 189

List of Figures and Table

FIGURES

Figure 1.1	The Dynamics of Culture-Place Domains	8
Figure 2.1	Food Insecurity in Las Vegas	34
Figure 2.2	Artist Rendering of VRCG Development Plans	35
Figure 4.1	Raised Bed Plots at VRCG	73
Figure 4.2	Randy's Three-bin Composter	75
Figure 4.3	Fence Mural	80
Figure 4.4	Fence Mural 2	81
Figure 4.5	Fence Mural 3	81
Figure 4.6	Fence Mural 4	82
Figure 4.7	Fence Mural 5	82

TABLE

Table A.1	List of Semistructures In-Depth Interviewees	168

Acknowledgments

This book would have been impossible had Rosalind "Roz" Brooks not welcomed me so generously at the garden. She warmly opened up the garden to me as a place to research, garden, and relax. I have innumerable fond memories of hanging out with Roz and other volunteers in the shade under the misters, by the fire drinking Miller High Life, at a picnic table watching the sunset, and in the rows discussing the crops. I spent a majority of my summers, and a great deal of my springs and falls, throughout graduate school at the garden. It became an important "third place" for me that helped me get through the stress of a PhD program. Roz's friendship and hospitality, along with that of many other volunteers, were largely responsible for the garden becoming a home away from home for me while I lived in Las Vegas.

Michael Borer was both a great mentor and friend throughout the entire process. He took a great interest in the garden and its relationship to the redevelopment efforts of the Downtown Project that was unfolding while I was in the field. We shared many hours discussing what I was observing at the garden and how to make sense of the patterns that emerged. I am so thankful for the numerous opportunities he provided, and continues to offer, to work together. Robert Futrell generously listened to me recount tales from the field, talked through puzzles and preliminary analysis, and encouraged me to pull on threads and pursue lines of inquiry, which helped me focus a research project that could have gone in countless different directions. His graduate seminars were the "seed plots" from which many of the theoretical and conceptual tools that came to form the backbone of this research were cultivated.

David Dickens and Simon Gottschalk provided extensive and thorough feedback to every chapter of this manuscript, which greatly enhanced the final document. Steve Clarke's expertise as a landscape architect and his experiences working in the Historic West Las Vegas neighborhood resulted in interesting and thought-provoking conversations about this research and pushed me to reconsider some of my observations and assumptions. I am also grateful for the comments, critiques, and suggestions made by the anonymous reviewer, as well as the guidance from my editor, Courtney Lachapelle Morales. Their feedback and support strengthened the final product.

I am also eternally grateful to my Mom and Dad, who have been nothing but supportive throughout my academic journey. They have cheered me on from a distance for every final exam, conference presentation, degree defense, and job interview. Unlike some families that never fully understand what their sociologist relatives do, my parents have always listened, asked questions, and taken a genuine interest in my work. I am truly a lucky person to have them.

Las Vegas would not have been the same had I not shared an adorable little pink house in the Huntridge neighborhood with my sister, Kerri, for most of my graduate studies. She helped me eat the food I brought home from the community garden and was always willing to let me bounce ideas off her as I interpreted what I was seeing at the garden. I feel extremely fortunate to have had the opportunity to live and learn with her as adults in Las Vegas.

Finally, I do not know how I would have finished this manuscript without the support of my partner, Aleta. Always willing to read drafts, talk through ideas, and offer up words of support and encouragement, she was instrumental in helping me get across the finish line. Having the unwavering belief of such a brilliant person behind me and my work is an incredible gift.

Chapter 1

Sowing Seeds of Community and Sustainability in a Wasteland

I did not set out to write the obituary of a community garden. About a year into my fieldwork, however, it seemed this was precisely what I was going to have to do. It appeared as though the harsh cultural and natural environments of Las Vegas were going to sabotage the garden before it could take root in the heart of the city. As the months passed, it was apparent that although the garden was financially struggling, and some plans were scrapped and volunteer turnover levels were high, there was an undeniably material survival occurring; the plants were still growing. Instead of a story of triumph or failure, this is about a community garden's *persistence* in Las Vegas.

Moving to Las Vegas from Central Missouri was a bit of an adjustment. After living in the city for six years, I came to understand this was true for transplants from nearly anywhere. Las Vegas just takes some getting used to. It took me about a year. I moved from a rather uninspiring gated apartment complex nestled right on the border between Las Vegas proper and Henderson, NV, to a small, but cute, older home in the historic Huntridge neighborhood in Downtown Las Vegas. The house had a backyard with two dirt patches that had clearly been used in the past for growing something. I decided I wanted to try growing vegetables. I never grew food in Missouri so I had no idea what I was doing, and I was especially intimidated by the prospect of keeping plants alive through a Mojave summer. I got online and found that Vegas Roots Community Garden[1] was located a few minutes' drive from my house and stopped by the next day.

I started going to the garden a few times a week and quickly began to feel a connection not only to the garden and the handful of volunteers with whom I was interacting but with Las Vegas itself. I felt connected to Downtown Las Vegas and to the Southwest. While working in the garden that first summer, I came to embrace Las Vegas' infamous "dry" heat. Perhaps it was the year

I had spent in the city that made me feel more at home or the friends I had made through UNLV, but it was also absolutely a result of my experiences and interactions at the garden. The garden and the people associated with it came to help me feel like a Las Vegan, and not just a graduate student passing through town. Building a relationship with the land itself helped me feel more firmly rooted in the city. I was no longer just looking at the highly stimulating city through my eyes, but instead experiencing it through all of my senses, not the least of which involved tasting it.

Sociologist and demographer Kingsley Davis claimed, "The *sine qua non* of urban existence [is] the concentration in one place of people who do not grow their own food."[2] In a sense, this is true. Technological advancements in agriculture played, and continue to play, a large role in global urbanization processes. As fewer people need to dedicate their lives to food production in order to feed the nation, many who would have previously gone into farming moved to cities in search of economic opportunities. Over time, food became an impersonal item purchased from, most often, a chain grocery store. Michael Pollan, the critic of industrial, global, processed food, stated, "70 years ago, everybody knew where their foods came from. You couldn't write a book about that. But now, nobody really has any idea."[3] So yes, Davis's statement, for the most part, accurately depicts the American urban experience of the past century: cities are for people, food comes from some other, often unknown, place.

Easily the most common reaction when I tell others that I study a community garden in Las Vegas is disbelief. This response comes from academics as well as many Las Vegas locals. When people think of Las Vegas, their thoughts are rarely accompanied by images of lush urban agriculture. As a tourist destination, Las Vegas is one of the most popular cities in the world. As a metropolitan area, the city is often perceived and discursively interpreted as marginalized. The strength of this impression has even led respected urban scholars to make the "stupefying"[4] claim that Las Vegas is "not a real city."[5] In large part, these ideas are a product of a "wasteland discourse" that surrounds the city. There is a perception that Las Vegas is both a cultural and an ecological wasteland. The city's reputation as an unsustainable theme park dominated by superficial, inauthentic culture leads many to the conclusion that Las Vegas residents must be blinded by the neon into submission to this empty culture instead of having meaningful urban experiences, interactions, and, ultimately, existences.

COMMUNITY FOUND IN A TRANSIENT CITY

"Community" is one of those concepts across academic disciplines and popular discourse that is infamous for its imprecision. The abundance and diversity of definitions are not indicative of a bad concept, though. Instead, it is a reflection of the complex messy ways people experience and use community.[6] Broadly speaking, however, definitions of community tend to involve three elements. First, some definitions emphasize *territoriality*, or the spatial proximity of community members, foregrounding the importance of shared space between community members. This is, in part, because many community studies are, and have been, neighborhood studies.[7] Second, there are definitions that focus on the *social ties or networks* that connect community members. The third element is less present in the literature, but Margarethe Kusenbach's work on "neighboring" is an exemplar of an approach that focuses on the *meaningful interactions* that serve as the glue holding together community members.[8] Calling attention to the ways in which community is practiced, Kusenbach focuses on patterns of "communal interaction," which she argues are "the normative, habitual practices that build and perpetuate communal relationships."[9]

For decades academics have argued over the alleged disappearance of community from American cities. Although scholars have made compelling claims that community has been "found," "saved," and "liberated," in many cities, Las Vegas residents report low levels of "neighborliness," and, predictably, tend to feel less "sense of belonging" to their neighborhoods than to the city, metro area, state, southwest region, and nation.[10] While Lyn Lofland describes the "public realm" of cities as a "world of strangers,"[11] in Las Vegas, this characterization also applies to much of the "parochial realm" of neighborhoods, creating obstacles for community formation.

Las Vegas has gained a reputation for a lack of community, and this reputation is at least in part a result of its status as a tourist destination. Many tourists only experience the city via The Strip or the Fremont Street Experience. Although they may see the sprawling metropolitan area as they fly in and out of the city, they often fail to consider how the lives of the roughly two million residents of the Las Vegas valley, including values, beliefs, desires, and needs, are much like their own. Among residents, though, who live, work, and play in the city, their experiences closely resemble the external reputation of the metro area. When talking to residents about varied social, economic, and environmental issues impacting the city, a common "frame" used to explain these issues is one that centers on some variation of the claim that "Las Vegas is a transient city." While "frames" can help social actors interpret stimuli, they can also become set paths for interpretation that come to serve as justifications for the status quo and lead actors to forego considerations of alternatives.[12]

Many are familiar with the superlative growth of Las Vegas in the 1990s and early 2000s. Indeed, the Las Vegas metropolitan area grew at a rate of 83 percent over this decade making it the fastest growing in the country. This growth continued for nearly another decade until the recession and the accompanying housing crisis hit the city. A lesser-known fact was the simultaneous high rates of exodus from the city. During the height of the city's population boom in the early 2000s, the population was growing by 6,000 residents per month, but roughly 9,000 residents were moving in. Accordingly, 3,000 residents were leaving Las Vegas during the same period.[13] This rapid turnover created neighborhoods with few long-term residents.

In 2009, right as the impacts of the Great Recession came into view, a team of UNLV sociologists collected survey and focus group data citywide on residents' knowledge and attitudes about social, economic, and environmental issues in the city. Their findings in the Las Vegas Metropolitan Area Social Survey (LVMASS) paint a picture of a city with a relatively unstable population. First, only 8 percent of respondents were born in Las Vegas. Second, the average time respondents had lived in their homes was only 12.1 years, and 34 percent of respondents reported having moved to their current homes from out of state, reflecting not only a high overall turnover rate in the city but also within neighborhoods. This shuffling of residents into, out of, and across the city contributes, at least in part, to a diminished sense of "place attachment," or the sentimental connection people establish with places rendered meaningful via social interaction, among residents.[14]

Of those surveyed by the LVMASS team, less than 37 percent reported feeling attachment to Las Vegas, and about 34 percent reported feeling attachment to their neighborhood. Additionally, the data suggested that those born in Nevada were more likely to feel attachment toward the city, although roughly 75 percent of respondents were born out of the state, and about 16 percent were born in other countries. Finally, the proportion of those who reported feeling a sense of belonging to Las Vegas or Nevada was higher than those reporting similar sentiments toward their neighborhoods. All of the aforementioned findings highlight Las Vegas residents' rather shallow roots in both the city and their neighborhoods, but the finding regarding low attachment to neighborhoods may also suggest weak "social capital" for those who were part of the recent population spike.

LVMASS data supports this. Among survey respondents, 41 percent said they "almost never" visit their neighbors, 63 percent said they "almost never" do or receive favors from neighbors. Focus group data echoed these findings with participants overwhelmingly reporting weak bonds with their neighbors and a sense that most people in their neighborhoods were strangers. Furthermore, many residents in the focus groups reported desiring more and stronger community bonds. These data imply a lack of social capital among Las Vegas residents in their neighborhoods. Briefly put, social capital refers to relationships and networks individuals can tap into in times of need, and that underlying trust that makes cooperation and coordination possible.[15]

These shared support structures are common elements to many scholars' conceptualizations of "community."[16]

There is a chance the recession has a silver lining for Las Vegas. The housing crisis and unemployment rates were devastating to Las Vegas residents, but these trends could slow the rates of population turnover that create the transient conditions characteristic of Las Vegas. That is, if people are forced to stay, at least for a little while, they may begin to invest more in the city and their communities.[17]

> When we cease to be migrants and become inhabitants, we might begin to pay enough heed and respect to where we are. By settling in, we have a chance of making a durable home for ourselves, our fellow creatures, and our descendants.[18]

This is a sentiment that circulated the Las Vegas valley in the aftermath of the housing crisis[19] and shared by the director of the community garden a few years ago when she changed the name from Tonopah Community Garden (it is located on Tonopah St.) to Vegas Roots Community Garden. She wanted to start a citywide initiative to convince people to "put down roots" in Las Vegas and start improving the city socially, economically, and environmentally.

Like its fellow Sunbelt cities, Las Vegas is a sprawling metro area characterized by land-use segregation and the pervasiveness of the somewhat oxymoronic gated development "communities." Gated neighborhoods and rapidly growing transient populations go hand in hand, as gates are often employed as security and safety mechanisms to protect interior residents from the uncertainty and perceived danger that accompany rapid demographic change and social change in general.[20] Indeed, the rhetoric typically used to claim gated developments provide the necessary conditions for fostering communal bonds often fail to pan out, instead offering only a perceived increase and safety and friendliness, and less often the communal participation associated with social capital.[21]

It was in this context that Vegas Roots Community Garden set out to forge community in Las Vegas. Throughout the remainder of this manuscript, I explore how community formation was strategized at the garden, what shape it took, and how to interpret and situate the resulting contradictions. For example, how did reliance on day laborers to clear the property to make space for the garden affect the meaning of community at the garden? Under what circumstance does borrowing ideas and verbiage from existing successful gardening projects help a new project formulate its own sense of purpose, and when does reliance on these conventions start to sabotage a new project? How do goals change over time, and how are these changes affected by the reception of the project? How do narratives created by members of the group shape the ways they interpret the reception of the project? These are some of the questions I explore and attempt to answer throughout the following chapters.

SUSTAINABILITY IN A CONSPICUOUSLY WASTEFUL CITY

Studying community gardening in Las Vegas also adds complexity to existing understandings of the practice because the city has a reputation for being both wasteful and for being a wasteland itself. Over 80 percent of Nevada is owned by the federal government, and Kuletz illustrates the ways in which vast swaths of the Mojave Desert in the state have become "sacrificial landscapes."[22] The federal government has socially constructed parts of the desert as uninhabitable wastelands in order to rationalize, for instance, their use to test above ground and underground nuclear weapons at the Nevada Test Site. More recently, the desert surrounding has been used to store low-level nuclear waste, and Yucca Mountain, just outside the city, has been ensnared in political debate as a potential long-term storage site for spent nuclear fuel rods from nuclear power plants. These materials remain highly radioactive for *thousands* of years.

This reputation is supported by relatively weak local activism to curb waste in the Las Vegas valley. There is little faith in the city's recycling program among locals, the highly developed bar and nightclub industry and casinos do little to mitigate the massive amounts of potentially recyclable waste they produce, and the city is known for its stubbornness about reusing or repurposing casinos, opting instead for highly publicized and televised explosions or implosions. Furthermore, although Southern Nevada has made significant strides in curtailing water waste in the face of the disappearing Lake Mead, Las Vegas' primary water supply, there are still large amounts of water wasted on landscaping in order to support plant life not native to the region. Mike Davis describes the latter phenomenon as "hydrofetishism" that ignores the hydrological realities of the desert Southwest.[23]

At a more basic level, the city is seen as wasteful in its promotion of gambling or coercing patrons into paying more than twenty times more than retail price for bottles of alcohol in nightclubs, each of which are viewed as "throwing away" money. Additionally, as one walks down Fremont Street under the "Viva Vision" canopy of the Fremont Street Experience in the summer heat, one notices the air conditioning from the open doors of the casinos control the outside climate, wasting the large amounts of energy to cool the massive casinos.

The garden is promoted as an alternative to the Las Vegas status quo. The director and numerous volunteers told me that part of the reason they enjoy spending time at the garden is that it does not feel like it is part of Las Vegas.

They believe the values and beliefs that undergird the garden are not just different than those they see as guiding the gaming and nightlife aspects of Las Vegas, but morally superior. Part of the garden's mission is to try to create a more sustainable food system. Part of this effort revolves around trying to convince residents to eat more locally grown food and to encourage behaviors like recycling and growing food at home. What exactly a sustainable food system would look like, what is unsustainable about the current food system, and the extent to which the garden itself *is* sustainable is never meaningfully addressed at the garden or in its promotional materials. These are issues I return to in subsequent chapters.

COMMUNITY AND CULTURE IN LAS VEGAS FROM THE GROUND UP

Place and Culture in the City

My research contributes to a growing body of work from an urban culturalist perspective (UCP hereafter).[24] Whereas most urban research starts with a research topic and then seeks out a place where that topic exists, the UCP urges researchers to begin with a place and ask a simple, yet broad question: *What happens here*? This theoretical approach to examining urban places and practices treats culture as more than a dependent variable, compared to other traditions where culture is understood as a by-product of political and economic forces. Unlike traditions that theorize culture as mostly structurally determined, the UCP emphasizes the ways in which everyday urbanites actively use both urban space to (re)construct shared systems of meaning and these collective cultural constructions to inject meaning and order into an otherwise largely anonymous, chaotic urban existence.[25]

Research in the UCP tradition is guided by the assumption that social actors are actively involved in the *creation* of culture. Urbanites are not simply the recipients or transmitters of culture imposed from the outside.[26] Urban places often play a critical role in the construction of culture by serving as resources facilitating meaning-making and influencing interactions and identities. Place refers to the physical sites that are rendered meaningful through interactions that occur in them.

Before going any further, it is important to specify what I mean by culture. Throughout this manuscript, I will be discussing and analyzing culture at two distinct but interrelated levels. First, at the micro, every day, interactional level, culture involves meanings, values, beliefs, and sentiments everyday social actors create to order social reality and render social life meaningful, *as well as actions based on these meanings*. As Kathleen Blee points out, "culture *is* meaningful action."[27] Second, at the macrolevel, culture is a space, or repository, where social actors can access collective meanings about what it means to be a member of a society, what values are shared among members, and the "moral order" of the group. Culture at the macrolevel can shape and constrain the actions taken by individuals on a microlevel. Culture at the microlevel can reaffirm, challenge, and ultimately change the macrolevel meanings and sentiments. Approaching culture at both levels simultaneously allows one to examine the ways in which culture is both *created* by and transmitted *through* individuals interacting in small groups. Moreover, using this conceptualization of culture, I will illustrate some dynamic processes through which collective actors simultaneously draw from macrolevel cultural trends and conventions and create place-based and place-specific meanings and narratives to make sense of their collective efforts to themselves and to the public. I fully acknowledge there are limits to social actors' abilities to have complete control over culture, and that these limits often follow familiar fault lines of privilege and inequality, meaning that with privilege comes increased productive power.

My work extends the UCP in two ways. First, by examining the processes through which social actors construct various domains related to a place, I highlight *how* these culture-place relationships are created. Second, through my case of Vegas Roots Community Garden, I illustrate some ways in which once created, these culture-place domains act back on the place itself and participants (see figure 1.1). This research adds complexity to the UCP by demonstrating how the creation of these culture-place domains through interaction and their subsequent use in interactions shape future actions in and on behalf of the place. One important way these domains impact action is in the

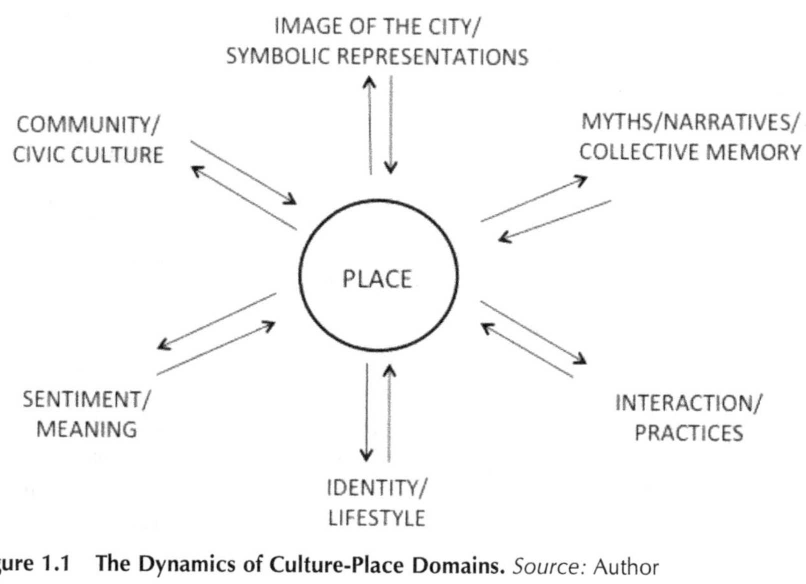

Figure 1.1 The Dynamics of Culture-Place Domains. *Source:* Author

ways they alter the meaning of what is possible, the boundaries of the imaginary, and the perceptions of potentials.

Agency and Collective Action in the City

In line with the assumption that individuals are not simply given culture, but instead aid in its construction, the approach to grassroots collective action taken here highlights the agency of social actors and small groups. Reflecting Marx's assertion that social actors exercise agency, but only under conditions over which they have little or no control, I examine the choices made, and those not made, when participants encounter obstacles and opportunities.

I also draw from the work of Mustafa Emirbayer and Ann Mische on agency and temporality, which is premised on the notion that agency does not only unfold *over* time but also *in* time. They suggest that agency is a complex process that at different times is oriented toward the past (drawing on habits), the future (aimed at goals), or the present (reflecting on current self-assessments). The temporal orientation that takes precedent in a particular situation is contextual.[28]

Sometimes the strategic choices collective actors make lead to what Blee calls "turning points." She posits that collective actors often develop routine ways of operating, or patterns of action, which emerge from choices that are typically future oriented. This sets them down paths or trajectories in which decisions about lines of action are perceived as given, often past-oriented based on habit or tradition. At certain points, these trajectories change course, sometimes based on present-focused assessments of the group, other times based on rethinking future-oriented group goals.[29] I elaborate much more on these issues of agency and strategy in chapter 5.

I examine various outcomes of the strategic choices garden participants make at the community garden, but one of the major dynamics I revisit throughout the book is how strategic choices continually affect the ability for participants to forge a sense of collective identity. Simply put, collective identity refers to "the processes through which a collective becomes a collective."[30] Collective identity formation does not occur by simply tapping into preexisting commonalities among collective actors. Instead, it is an ongoing process of negotiation and co-construction of a bond that helps collective actors make sense of what they are doing together.[31]

Nature in the City

Another theme that permeates throughout the following chapters is the role of nature in cities. This is a topic that has recently gained increasing attention in sociology in various different forms, but which has been present in other disciplines such as geography, environmental psychology, and architecture for quite some time. Sociologists are interested in the impacts of natural features on urban residents' quality of life, the role of natural features on urbanites' perceptions of city spaces, and the ways in which the types of natural features in a given urban space reflect cultural meanings, beliefs, and ideals.

At a basic level, biophilia refers to the "love of life."[32] Since Erich Fromm's introduction of biophilia in psychology, biologist E.O. Wilson expanded on the concept to develop the "biophilia hypothesis," which posits that there is an instinctive, evolutionary bond between humans and other biophysical beings and organic matter.[33] Many of my interview respondents reported experiencing a multisensory pleasure from simply putting their hands into soil. Indeed, neuroscience research suggests exposure to bacteria found in soils can increase serotonin levels in the brain, reducing depression and anxiety.[34]

Recent research has documented numerous ways in which city dwellers experience psychological, chemical, and physical changes when exposed to nature in their everyday lives. The positive effects documented in this research are so wide reaching that some scholars now refer to green spaces, or urban spaces containing concentrated natural features such as trees, animals, and/or water, as "Vitamin G."[35] Although one might attribute the term "aromatherapy" to a savvy candle or diffuser marketer, some data suggests that olfactory exposure to flowers and other plants can reduce stress and even reduce health recovery times.[36] Research also indicates walking through natural settings can improve memory and attention in future activities.[37] Furthermore, some data suggests that simply looking at images of nature can improve cognitive performance and mood,[38] and indoor plants can alter the mood of a workplace and reduce stress.[39]

The data suggests natural elements improve quality of life in urban settings. This growing body of literature has begun to inform the ways that some urban planners, designers, landscape architects, and others think about how to make cities better for residents. Some sustainability scholars argue that the real and perceived restorative effects of human contact with nature has, in part, driven the desire for suburban living which ushered in land use and transportation infrastructure that challenges sustainability and that urban planning that includes green space could allow city dwellers to achieve the restorative qualities of interacting with nature without leaving the city, encouraging more sustainable residential patterns.[40]

Finally, proximity and views of green spaces have been associated with lower crime rates,[41] increased "life satisfaction,"[42] heightened neighborhood satisfaction,[43] and heightened prosocial and community-oriented behaviors.[44] Whereas some of the previous research results I identified may indirectly impact those around the individual exposed to nature, these effects of nature are purported to be more social and impact those in proximity, whether they engage with the green space or not.

How these findings directly translate to specific types of green spaces like community gardens is an empirical question. Furthermore, it is critical to remember that what constitutes nature is culturally negotiated and what nature means in a given place or situation is also culturally contextual.[45] Therefore, I draw from scholars and activists who critique the contexts and politics of human-nature interactions in cities.

Gardening and Justice in the City

Throughout the book, my analysis is also informed by the interdisciplinary body of literature on urban gardening. The research on nature and green space in cities overlaps with that on urban community gardening. There is a great deal of literature exploring the extent to which community gardens create the benefits described above for those who participate in such projects as well as additional positive social, physical, and mental outcomes.

Research indicates that volunteers at community gardens practice and reproduce democratic values within the garden space.[46] Although the presence of a community garden in a neighborhood does not necessarily reduce crime rates, it can have the impact of altering residents' perceptions of safety, their investment in the neighborhood, and their health behaviors at home.[47] They can also serve as sites to discuss community issues and plan social action to address them.[48]

Previous research finds that garden participants identified community gardens as places to meet and interact with people for varying cultural backgrounds. They are sites that provide occasions for interactions between different genders that challenge traditional gender roles,[49] and unique opportunities for interracial interaction.[50] Since participation in voluntary associations is linked to social capital development,[51] it is no surprise community garden participation has also been shown to produce mutual trust, reciprocity, social support,[52] social cohesion, social connections, cooperation, bonding, bridging (i.e., "links that cut across various lines of social cleavage"[53]), but maybe limited to the physical space of the garden.[54]

Aside from social benefits, community gardens and community garden*ing* are often celebrated for producing psychological benefits. Community gardening can increase participants' self-worth, self-confidence, and self-respect,[55] whether it comes from communal endeavors such as collaborative gardening, event planning, design input, and beautification of the space, or from more individualized actions like successfully tending a plot. The work of gardening promotes physical activity, which even at moderate levels is associated with improved mental and physical health.[56] Gardening can also make people feel good, or at least OK, because it represents an aspect of their lives they can control. This is a point echoed by Nelson Mandela:

> A garden was one of the few things in prison that one could control. To plant a seed, watch it grow, to tend it then harvest it, offered a simple but enduring satisfaction. The sense of being the custodian of this small patch of earth offered a taste of freedom.[57]

The sense of control granted by the cultivation of plants has helped other groups experiencing trauma and crisis throughout history.[58] Helphand described soldiers engaged in trench warfare in World War I, Jews stuck in Warsaw ghettos, and Japanese Americans imprisoned in World War II internment camps who turned to gardening for relief.[59] He argued these were not attempts to escape these inhumane conditions, but instead acts of resistance that granted individuals some degree of control amid a great deal of chaos.

Mandela's and Helphand's insights are not just relevant to the imprisoned and those engaged in war; in cities they are especially true for poor, elderly, physically or mentally handicapped, and the disenfranchised.[60] Indeed, for the elderly, communal gardening provides opportunities to combat isolation,[61] gives them something habitual to look forward to, and, again exercise control over an aspect of their lives at a time when that is likely something they are losing.[62] Community gardens are also regarded as critical sites for relaxation and respite amid an otherwise hurried, hectic, and at times hostile urban environment.[63]

It is important not to forget the dominant activity in community gardens, though: growing food in the city. While all these physical, social, and psychological benefits are potentially fostered through participation at community gardens, they are often understood as positive side effects of, and often justifications for, creating alternative, grassroots food sources in cities. In addition to being promoted on the basis of their positive contributions as urban green spaces, community gardens are often held up as one of the main types of spaces in the movement toward more sustainable alternatives to the industrial, global food system.

In order to grapple with this second dimension of community gardening, I draw on literature from the alternative food and food justice movements. I do this for two reasons. First, garden participants use the language of these movements in promotional materials as well as in conversation. Importantly, not all participants use this language, and even when it is employed, it is often done so abstractly. Second, although community gardens are created for a diversity of reasons, one motivation that is commonly cited is a desire to promote alternative food and food justice.

The alternative food movement, sometimes simply referred to as the food movement, is primarily focused on opposing and creating alternatives to the increasingly industrialized, rationalized, and globalized food industry created to feed an expanding human population and maximize profits for the increasingly consolidated and corporatized food producers. The industrialization of food production has resulted in the rise of mechanized farming practices and a reliance on monocultures—food production practices that favor the cultivation of homogenous, genetically engineered crops, requiring heightened amounts of chemical pesticides and water. The current configuration of mechanized global food production has created high demands for fossil fuels, both by utilizing large-scale planting, cultivation, and processing machinery that is highly energy intensive as well as the increasing distance food travels from farm to fork,[64] most of which is made possible by fossil fuels.

Alternative food concerns center around issues of locality, safety, and environmental sustainability. At one point when the director of Vegas Roots Community Garden was discussing the possibility of starting a pick-it-yourself farmers market at the garden, she excitedly commented, "It doesn't get any more local than that!" She was referring to the fact that there are farmers markets in Downtown Las Vegas, but much of the produce sold at these markets come from farms outside the city or even from Southern California. A market at the garden that offered food *grown* in Downtown Las Vegas would be *more* local.

Some of the alternatives proposed by this movement include creating new spaces of food production—like small farms in the urban hinterlands, community, and other urban gardens and farms, and rooftop gardens—and the organization of these spaces into networks so that others are able to purchase and consume food cultivated outside the industrial food system. These networks are what make many urban farmers markets possible. Activists often refer to those who opt into these alternative networks as "voting with their forks," that is, making a political decision to "invest" in alternative economies, which also necessarily entails divesting in the status quo.

More recently, activists and scholars of alternative food have begun to look more critically and reflexively at the narratives and practices, as well as demographics, of the movement. Many who do so argue for a shift in both the academy and in practice toward addressing problems in our global food system through approaches rooted in "food justice." When I use the term "food" justice, I refer to the body of literature and social activism that has sought and continues to seek to expand the mission of "alternative" food movements to address the various ways social class, race and ethnicity, and other dimensions of privilege and oppression impact the ways in which the environmental "goods" and the "bads" of food production, distribution, and consumption in American cities. Food justice, then, is a perspective that combines social justice, environmental justice, and alternative food.

Some of the major insights from this line of inquiry have been an acknowledgment that much of the politics of alternative food have been undergirded with color-blind and universalist ideologies.[65] Julie Guthman examines the phrase "if they only knew . . ." in order to demonstrate how narratives operate at both macro- and microcultural levels in ways that tend to reflect the experiences and interests of privileged status groups. "If they only knew . . ." is a narrative that suggests if people knew more about the conditions in which food is produced in the age of industrial food production, many would "vote with their fork" for transformation of food systems toward more ethical and sustainable practices. This narrative operates at the macrolevel via dominant cultural exemplars of the alternative food movement like Michael Pollan and Eric Schlosser and their popular press best sellers,[66] films like *Forks Over Knives*,[67] and literary explorations of industrial animal production like Johnathan Safran Foer's *Eating Animals*.[68] While these, and many other accounts of our food system are not untrue, as Guthman points out, they present the problem in a way that implies a belief that

> an unveiling of the American food supply would necessarily trigger a desire for local, organic food and people would be willing to pay for it (cf. DuPuis 2001). Then, so the logic goes, the food system would be magically transformed into one that is ecologically sustainable and socially just.[69]

The bottom line for critics is that these approaches are premised on the assumption that knowledge produces action, which sparks change. Furthermore, this message, which often romanticizes and calls for a return to past agricultural traditions and neglects to acknowledge the highly racialized aspects of both past agricultural traditions and the current food system, has been formulated largely by privileged individuals based on historically privileged relationships to the industrial food system. Accordingly, the alternative food movement has created a message that largely resonates with middle- and upper-class whites.

On the microlevel, there is a tendency for alternative food spaces, those locations that, informed by the messages of the food movement, try to provide alternatives to industrial agriculture end up being "coded" as white spaces. That is, their foundation in a narrative that misrepresents or completely ignores the racial inequalities in past and present food systems discourages racial and ethnic minority participation, which could potentially foster more comprehensive objections and alternatives to the status quo. Some of the omissions food justice advocates highlight are the connections between critiques of the food industry that focus primarily on the compromised nutritional value and taste of industrial food without critiquing the working conditions of the laborers who prop up the industry. Also missing from the dominant food movement's messaging is a nuanced critique that acknowledges the unique ways in which the industrial food system continues to produce hunger and food insecurity in poor regions and neighborhoods.

Scholarship on gardens and gardening have similarly tended to romanticize these spaces and practices, neglecting to acknowledge their racialized and class-based characteristics. Sociologist Pierrette Hondagneu-Sotelo poses the poignant question, "When do we experience gardening as pleasurable, and when do we regard gardening as a dreaded chore that should be outsourced if resources allow?"[70] As this quote highlights, scholars and activists are quick to elevate the spiritual, transcendent, and restorative elements of gardening, but what about when gardening is work and not simply a leisure practice? How do experiences of poor, immigrant landscapers (gardeners) or migrant farm workers (gardeners) trouble what we know about experiences in gardens and in nature? It is in these theoretically murky spaces where I will grapple with what it means that a community garden periodically relies on day laborers to complete crucial projects and maintain the visual aesthetic of the garden.

WHY STUDY GARDENS?

Making a case for increased sociological attention to gardening, Pierette Hondagneu-Sotelo points out, "Gardens are places where nature and culture meet."[71] Accordingly, studying a community garden in Las Vegas provides an entry point into understanding the dominant culture of the city and how residents use and repurpose it. Additionally, it can provide a window into understanding Las Vegan's complicated relationship to the local ecosystem. By analyzing the trials and triumphs of a community garden project in Las Vegas, I am able to highlight the ways in which ideas, both from outside and within the city, about Las Vegas's cultural and ecological landscapes influence efforts at community building and collective action.

Gardens are not simply "nature," since notions about what constitutes nature are socially constructed and therefore reflect the cultural contexts in which they are located.[72] Furthermore, gardens are fundamentally about the *control* of nature. There is, however, a material reality to the natural aspects of a garden, even when located in the "concrete jungles" of American cities. For instance, there are ecological realities of geographic regions with which a gardener must contend. The American Southwest is 19 years into a megadrought more severe than any of the same duration in centuries,[73] thus forcing gardeners to adapt by either growing less water-intensive plants or relying more on municipal water. Growing seasons also vary widely across the United States, limiting when one can grow, and what will grow at different times of year. Soil qualities also vary regionally, something many Las Vegans learn the first time they try to penetrate the soil and find the rock-solid caliche clay underneath thin layers of topsoil. Finally, even though gardening involves a degree of domination over nature, sometimes the agentic choices made in a garden trigger natural responses, like the introduction of weeds, an increase in bee populations, an increase in pests, and alterations in pollen profiles (to name but a few). Accordingly, urban community gardens are locations in which one can examine the intersecting importance of place, culture, and nature in city life.

One of the major tensions identified by sociologists with regard to urban gardens and other natural features in cities involves the questions of who gets to enjoy these spaces and who must labor to maintain them and under what conditions.[74] This illuminates the fact that urban gardens, whether at private residences or in "public" spaces are locations where inequality and power are at play.

Especially in the American Southwest, the role played by Latino immigrant gardeners in maintaining suburban yards and natural landscapes is highly visible. I was shocked the first day I visited Vegas Roots Community Garden (VRCG) and found Latino day laborers performing the difficult tasks of weeding the in-ground rows. I was even more surprised to later find out that day laborers were also responsible for the initial clearing from the property of old construction site debris to make way for the first raised beds.

One of the reasons it is critical to study gardens is their pervasiveness in everyday life. They are cultural objects that are present in some of our most private spaces and that facilitate highly sensual, sentimental, and even spiritual experiences for many people. Dating back to Thomas Jefferson's Monticello, home gardens have held a special place in the history of American domestic life. South of the Mason-Dixon line African slaves also attached a great deal of sentiment to the small plots of land outside their living quarters upon which they could exercise some semblance of autonomy in the cultivation of food.

Although important, the focus on the positive social, psychological, and spiritual impacts of time spent in gardens and other natural spaces has served to eclipse issued of power, privilege, and inequality.

THE LAY OF THE LAND

A central claim I make throughout this book is that the *hows* of community gardening in particular, and the practice of community more generally, are as important as the *whats*. No two food deserts are the same. Food insecurity is multifaceted. In chapter 2, I describe the context of my research in detail. I tell the story of the Historic Black Westside, a Las Vegas neighborhood with a history of structural racism and neglect that reaches back to the Jim Crow era and extends up to the present. VRCG is located at the edge of this neighborhood, and as such it is important to understand the historical, social, political, economic, and cultural backdrops against which the case at the heart of this research emerged.

In chapter 3, I examine the process through which participants constructed a religious narrative about the garden and came to interpret the garden as a sacred space. I illustrate the ways in which the anointment of the garden as an exceptional place distinct from the profane, mundane urban space that covers the majority of the city affects the ways participants understand collective efficacy, identify goals, envision paths for pursuing goals, and interpret their collective efforts.

In chapter 4, I analyze VRCG as a cultural object and explore the process through which garden participants produce it. In doing so I examine how VRCG participants appropriate conventions from the broader cultural trend of urban gardening in order to create a community garden in a local setting. This chapter expands on Howard Becker's (1974) work on cultural production but suggests that although reliance on cultural conventions can benefit cultural producers by saving time and facilitating interpretability among audiences, *over*reliance on conventions can be detrimental to both producers and the product. Additionally, the analysis in this chapter directly addresses the back-and-forth between macrolevel culture and the on-the-ground microelements of culture described above.

In chapter 5, I explore garden participants' strategic choices regarding organizational and interpersonal norms and goals. I explore how decisions made early in the garden's existence set the project down a particular trajectory making some actions more likely than others and rendering others outside the realm of possibility. This chapter also illustrates certain "turning point" moments in the garden's existence when participants reflexively analyzed the trajectory or rethought the goals of the project and how they came to either change trajectories or stay the course.

In chapter 6, I examine how issues of race, class, and geography are elided in participants' imagining of what community means at the garden and the impact that has on actions taken and assessments of progress and success. This chapter directly addresses the reciprocity built into the UCP that I illuminate in this research. The meanings of community at the garden are created by participants and eventually come to shape the imaginations of participants regarding what goals are possible, which ones should be pursued, and the extent to which the garden is "successful." Specifically, in this chapter, I analyze how not talking about race, class, and geography at the garden led to a conceptualization of community that was so inclusive that it created a scenario in which the garden was for everyone in Las Vegas. My argument in this chapter is that by trying to appeal to everyone, the garden was for *no* one.

In chapter 7, I examine the recent changes in the Historic Westside neighborhood and how they may impact the garden. I then discuss what the findings from the four preceding chapters can tell us more broadly about how collective actors in cities draw from broader cultural milieus to construct place-based culture, and how those creations act back upon the people who created them and the places in and about which they were created. In doing so, I also elaborate on how perceptual blind spots and contradictions are able to emerge and persist within urban green space-based collective action and cultural production and regarding social inequality and justice.

NOTES

1. At the time, the garden was actually still named Tonopah Community Garden.

2. Davis, Kingsley, "The Origins and Growth of Urbanization in the World." *American Journal of Sociology* 60, no. 5 (1955): 429–37.

3. Lauren Pantin, "Chatting with Michael Pollan of The Omnivore's Dilemma," *Teen Vogue*, September 28, 2009, http://michaelpollan.com/interviews/chatting-with-michael-pollan-of-the-omnivores-dilemma/.

4. David Dickens, "Is Las Vegas a 'Real City'?" *UNLV Gaming Research & Review Journal* 15, no. 2 (2012), http://digitalscholarship.unlv.edu/grrj/vol15/iss2/11.

5. In a video produced by Norton Publishing during the 2011 American Sociological Association meetings in Las Vegas, urban sociologist Sharon Zukin began her segment declaring, "I really hate Las Vegas." She went on to explain that people in the city only *think* they are having fun, but really they are just wandering through "inauthentic landscapes" in a "theater of consumption." Like many before her, her analysis of the metropolitan area and its roughly two million residents was based on limited exposure to the four-mile stretch known as The Strip. Although surprising from an academic, her willingness to dismiss the entire city as not "real" is illustrative of Las Vegas' position in the cultural consciousness.

6. Claire Nettle, *Community Gardening as Social Action* (New York: Routledge, 2014).

7. Barry Wellman and Barry Leighton, "Networks, Neighborhoods, and Communities Approaches to the Study of the Community Question," *Urban Affairs Review* 14, no. 3 (1979): 363–90.

8. Margarethe Kusenbach, "Patterns of Neighboring: Practicing Community in the Parochial Realm," *Symbolic Interaction* 29, no. 3 (2006): 279–306.

9. Kusenbach, "Patterns of Neighboring," 301.

10. Robert Futrell et al., "Las Vegas Metropolitan Area Social Survey 2010 Highlights," 2010, http://digitalscholarship.unlv.edu/reports/2.

11. Lyn H. Lofland, *The Public Realm: Exploring the City's Quintessential Social Territory* (Hawthorne, NY: Aldine de Gruyter, 1998).

12. Eduardo Bonilla-Silva, *Racism without Racists: Color-Blind Racism and the Persistence of Racial Inequality in the United States* (Lanham, MD: Rowman & Littlefield Publishers, 2003).

13. J. Patrick Coolican, "We're Stuck, So Will We Bond?—Las Vegas Sun News," *Las Vegas Sun*, December 25, 2010, http://lasvegassun.com/news/2010/dec/25/were-stuck-so-will-we-bond/.

14. Melinda J. Milligan, "Interactional Past and Potential: The Social Construction of Place Attachment," *Symbolic Interaction* 21, no. 1 (1998): 1–33.

15. Robert Putnam, *Bowling Alone: The Collapse and Revival of American Community* (Simon & Schuster, 2000).

16. Robert Ezra Park, "Human Ecology," *American Journal of Sociology* 42, no. 1 (1936): 1–15; Larry Lyon, *The Community in Urban Society* (Prospect Heights, IL: Waveland Press Inc, 1987); Edward James Blakely and Mary Gail Snyder, *Fortress America: Gated Communities in the United States* (Brookings Institution Press, 1997).

17. Coolican, "We're Stuck, so Will We Bond?"

18. Scott R. Sanders, *Staying Put: Making a Home in a Restless World* (Beacon Press, 1994).

19. Coolican, "We're Stuck, So Will We Bond?"

20. Blakely and Snyder, *Fortress America.*

21. Blakely and Snyder, *Fortress America*; Georjeanna Wilson-Doenges, "An Exploration of Sense of Community and Fear of Crime in Gated Communities," *Environment and Behavior* 32, no. 5 (2000): 597–611.

22. Valerie Kuletz, *The Tainted Desert: Environmental Ruin in the American West* (Routledge, 1998).

23. Mike Davis, *Dead Cities: And Other Tales* (New York: The New Press, 2003).

24. Michael I. Borer, "The Location of Culture: The Urban Culturalist Perspective," *City & Community* 5, no. 2 (2006): 173–97.

25. Borer, "The Location of Culture."

26. Borer, "The Location of Culture."

27. Kathleen M. Blee, *Democracy in the Making: How Activist Groups Form* (New York: Oxford University Press, 2012).

28. Mustafa Emirbayer and Ann Mische, "What Is Agency?" *American Journal of Sociology* 103, no. 4 (1998): 962–1023.

29. Blee, *Democracy in the Making.*

30. Alberto Melucci, *Challenging Codes: Collective Action in the Information Age* (Cambridge University Press, 1996).

31. Blee, *Democracy in the Making*; Steven M. Buechler, *Social Movements in Advanced Capitalism: The Political Economy and Cultural Construction of Social Activism* (New York: Oxford University Press, 1999); Melucci, *Challenging Codes.*

32. Erich Fromm, *The Heart of Man: Its Genius for Good and Evil* (New York: Harper & Row, 1964).

33. Edward O. Wilson, *Biophilia* (Harvard University Press, 1984).

34. Christopher A. Lowry et al., "Identification of an Immune-Responsive Mesolimbocortical Serotonergic System: Potential Role in Regulation of Emotional Behavior," *Neuroscience* 146, no. 2 (2007): 756–72.

35. Peter P. Groenewegen et al., "Vitamin G: Effects of Green Space on Health, Well-Being, and Social Safety," *BMC Public Health* 6 (2006): 149.

36. Seong-Hyun Park and Richard H. Mattson, "Therapeutic Influences of Plants in Hospital Rooms on Surgical Recovery," *HortScience* 44, no. 1 (2009): 102–05.

37. Marc G. Berman et al., "Interacting with Nature Improves Cognition and Affect for Individuals with Depression," *Journal of Affective Disorders* 140, no. 3 (2012): 300–05.

38. Marc G. Berman et al., "The Perception of Naturalness Correlates with Low-Level Visual Features of Environmental Scenes," *PLoS One* 9, no. 12 (2014): e114572; Netta Weinstein et al., "Can Nature Make Us More Caring? Effects of Immersion in Nature on Intrinsic Aspirations and Generosity," *Personality and Social Psychology Bulletin* 35, no. 10 (2009): 1315–29.

39. Virginia I. Lohr et al., "Interior Plants May Improve Worker Productivity and Reduce Stress in a Windowless Environment," *Journal of Environmental Horticulture* 14 (1996): 97–100.

40. Agnes E. Van Den Berg et al., "Preference for Nature in Urbanized Societies: Stress, Restoration, and the Pursuit of Sustainability," *Journal of Social Issues* 63, no. 1 (2007): 79–96.

41. Frances E. Kuo and William C. Sullivan, "Environment and Crime in the Inner City Does Vegetation Reduce Crime?" *Environment and Behavior* 33, no. 3 (2001): 343–67.

42. Mathew P. White et al., "Would You Be Happier Living in a Greener Urban Area? A Fixed-Effects Analysis of Panel Data," *Psychological Science* 24, no. 6 (2013): 920–28.

43. Rachel Kaplan, "The Nature of the View from Home Psychological Benefits," *Environment and Behavior* 33, no. 4 (2001): 507–42; Anne R. Kearney, "Residential Development Patterns and Neighborhood Satisfaction Impacts of Density and Nearby Nature," *Environment and Behavior* 38, no. 1 (2006): 112–39; Byoung-Suk Kweon et al., "Landscape Components, Land Use, and Neighborhood Satisfaction," *Environment and Planning B: Planning and Design* 37, no. 3 (2010): 500–17.

44. Nicolas Guéguen and Jordy Stefan, "'Green Altruism' Short Immersion in Natural Green Environments and Helping Behavior," *Environment and Behavior*, 48, no. 2 (2014): 324–42; Weinstein et al., "Can Nature Make Us More Caring?"

45. William Cronon, "The Trouble with Wilderness; or, Getting Back to the Wrong Nature," in *Uncommon Ground: Rethinking the Human Place in Nature*, ed. William Cronon (New York: W. W. Norton & Company, 1995): 69–90; Gary Alan Fine, *Morel Tales: The Culture of Mushrooming* (Urbana: University of Illinois Press, 2003).

46. By "democratic values," the authors are referring to the belief that people should actively participate in community politics beyond voting. Troy D. Glover, Kimberly J. Shinew, and Diana C. Parry, "Association, Sociability, and Civic Culture: The Democratic Effect of Community Gardening," *Leisure Sciences* 27, no. 1 (2005): 75–92; Michael S. Jamison, "The Joys of Gardening: Collectivist and Bureaucratic Cultures in Conflict," *The Sociological Quarterly* 26, no. 4 (1985): 473–90; Ellen Teig et al., "Collective Efficacy in Denver, Colorado: Strengthening Neighborhoods and Health through Community Gardens," *Health & Place* 15, no. 4 (2009): 1115–22.

47. M. R. Gorham et al., "The Impact of Community Gardens on Numbers of Property Crimes in Urban Houston," *HortTechnology* 19, no. 2 (2009): 291–96.

48. Teig et al., "Collective Efficacy in Denver, Colorado."

49. Diana C. Parry et al., "'Mary, Mary Quite Contrary, How Does Your Garden Grow?': Examining Gender Roles and Relations in Community Gardens," *Leisure Studies* 24, no. 2 (2005): 177–92.

50. Kimberly J. Shinew et al., "Leisure Spaces as Potential Sites for Interracial Interaction: Community Gardens in Urban Areas," *Journal of Leisure Research* 36, no. 3 (2004): 336–55.

51. John L. Hemingway, "Leisure, Social Capital, and Democratic Citizenship," *Journal of Leisure Research* 31, no. 2 (1999): 150–65.

52. Teig et al., "Collective Efficacy in Denver, Colorado."

53. Putnam, *Bowling Alone: The Collapse and Revival of American Community.*

54. Jonathan 'Yotti' Kingsley and Mardie Townsend, "'Dig In' to Social Capital: Community Gardens as Mechanisms for Growing Urban Social Connectedness," *Urban Policy and Research* 24, no. 4 (2006): 525–37.

55. Jamison, "The Joys of Gardening."; Tina M. Waliczekz et al., "Benefits of Community Gardening on Quality-of-Life Issues," *Journal of Environmental Horticulture* 14, no. 4 (1996): 194–98.

56. Masashi Soga et al., "Gardening Is Beneficial for Health: A Meta-Analysis," *Preventive Medicine Reports* 5 (2017): 92–99.

57. Nelson Mandela, *Long Walk to Freedom: The Autobiography of Nelson Mandela* (Boston, MA: Back Bay Books, 1994).

58. Pierrette Hondagneu-Sotelo, *Paradise Transplanted: Migration and the Making of California Gardens* (Berkeley: University of California Press, 2014).

59. Kenneth I. Helphand, *Defiant Gardens: Making Gardens in Wartime* (San Antonio, TX: Trinity University Press, 2006).

60. Jamison, "The Joys of Gardening."

61. Christina M. Gigliotti et al., "Harvesting Health: Effects of Three Types of Horticultural Therapy Activities for Persons with Dementia," *Dementia* 3, no. 2 (2004): 161–80; Christine Milligan et al., "'Cultivating Health': Therapeutic Landscapes and Older People in Northern England," *Social Science & Medicine* 58, no. 9 (2004): 1781–93.

62. Milligan et al., "'Cultivating Health.'"

63. Kate H. Brown and Andrew L. Jameton, "Public Health Implications of Urban Agriculture," *Journal of Public Health Policy* 21, no. 1 (2000): 20–39; Jonathan

'Yotti' Kingsley, Mardie Townsend, and Claire Henderson-Wilson, "Cultivating Health and Wellbeing: Members' Perceptions of the Health Benefits of a Port Melbourne Community Garden," *Leisure Studies* 28, no. 2 (2009): 207–19.

64. The data suggests that industrially produced food items travel more than 5,000 miles on average before finding itself on a plate (Weber and Matthews 2008).

65. Julie Guthman, "'If They Only Knew': Color Blindness and Universalism in California Alternative Food Institutions," *The Professional Geographer* 60, no. 3 (2008): 387–97.

66. Pollan, *In Defense of Food*; Schlosser, *Fast Food Nation*, respectively

67. Lee Fulkerson, *Forks Over Knives* (Virgil Films & Entertainment, 2011).

68. Jonathan Safran Foer, *Eating Animals* (New York: Back Bay Books, 2010).

69. Guthman, "'If They Only Knew'"; E. Melanie Dupuis, *Nature's Perfect Food* (New York: NYU Press, 2001).

70. Pierrette Hondagneu-Sotelo, "Cultivating Questions for a Sociology of Gardens," *Journal of Contemporary Ethnography* 39, no. 5 (2010): 498–516.

71. Hondagneu-Sotelo, "Cultivating Questions for a Sociology of Gardens." p. 511.

72. Cronon, "The Trouble with Wilderness; or, Getting Back to the Wrong Nature"; William R. Freudenburg et al., "Beyond the Nature/Society Divide: Learning to Think about a Mountain," *Sociological Forum* 10, no. 3 (1995): 361–92.

73. A. Park Williams et al., "Large Contribution from Anthropogenic Warming to an Emerging North American Megadrought," *Science* 368, no. 6488 (2020): 314–18.

74. Hondagneu-Sotelo, "Cultivating Questions for a Sociology of Gardens."

Chapter 2

"We Have a Community Garden in Las Vegas?"

Struggling with Segregation and Sustainability in Sin City

Roz[1] sat in a plastic chair behind a folding card table adorned with flyers and pamphlets with information about the community garden, fresh herbs and produce from the garden, and a trifold poster board, like the ones one might find at a school science fair, featuring pictures of the garden. It was a warm evening in March 2013 at First Friday in Downtown Las Vegas, a monthly festival celebrating local culture—art, food, entertainment, and, of course, booze. I asked Roz how the evening had been going so far. Exasperated and somewhat distressed, she said she could not believe how so many people in Las Vegas still had not heard of the garden. It had been open and operating for three years. She had said the same thing to me at an event in the summer of 2011, a year and a half after opening. She was not only frustrated that word was traveling so slowly but also optimistic because many who reported not having heard of the garden were excited to find out it existed. It also meant the lull in volunteers they were facing heading into the summer of 2013 could potentially turn around if they managed to mobilize the untapped pool of volunteers who had not heard of the garden yet.

Many people were not only unaware that Vegas Roots Community Garden (VRCG)[2] existed, they were surprised that there were *any* community gardens in Las Vegas. I discussed some of the cultural representations and reputations of Las Vegas in chapter 1, but residents' surprise regarding urban gardening in the city was, from my point of view, both surprising and completely predictable. On the one hand, residents have simply come to not expect such an activity in Las Vegas. Why is this the case? What has led them to make such an assumption? Is it unfounded? Yes and no. In this chapter, I will describe the research setting and context. It is an historical, political, and cultural excursion into the past and spanning to the present moment.

THE CITY

In order to understand many of the social, cultural, political, and ecological realities of present-day Las Vegas, one must consider some major, but relatively recent, historical developments in Southern Nevada. One hundred years ago, Las Vegas was a city with roughly 1,000 residents. It was little more than a train depot on the way to California and a small cluster of Mormons who had fled Utah. It was a desert settlement that exemplified the mythic images of the Wild West. It was rural. Shortly thereafter, in 1922, the lower Great Basin states (Arizona, California, Nevada, and part of New Mexico) decided to legally codify annual allocations to the Colorado River, an important source of water for each state. The members of the Colorado River Commission made these decisions based on their estimates of future water demands and allocated Nevada 300,000 acre-feet per year (AFY)[3] compared to 3.9 million AFY to California. Obviously, they were unable to foresee the growth Southern Nevada would experience over the next century.

Soon after the allocations were finalized in the Colorado River Compact, Nevada experienced growth in both the economy and population. This growth was only possible due to capital investments from the federal government. Throughout the 1930s, the government poured millions into the construction of the Hoover Dam, infrastructural improvements to Las Vegas, a U.S. Army gunnery range (what is now Nellis Air Force Base), Basic Magnesium chemical plant, and housing for the newly relocated defense workers.[4] By 1940, the city had more than 8,000 residents and the population of Clark County was twice as big. Moreover, the dam drew hundreds of thousands of tourists per year during and after its construction who stimulated the economy in Las Vegas by patronizing its recently legalized casinos.

The postwar affluence of many Americans provided increased opportunities for both recreation and residential mobility. Las Vegas' rapidly growing gambling economy provided pull factors not only for tourists but also people interested in working in the new casinos. The capital to finance the construction and operations at these new properties did not come from the federal government, though, but instead from organized crime. Throughout the 1950s and much of the 1960s the desire for economic growth in Las Vegas prompted most in the region to turn a blind eye to the criminal element behind the growth of gambling. After all, the Cultural Revolution of the 1960s resulted in altered cultural mores in America that embraced more self-indulgent leisure like gambling, drinking, and forms of entertainment offered in Las Vegas that was previous considered crude. By 1967, the population of the Las Vegas valley had reached 269,000 residents, and gambling was the primary economic force.

By the early 1970s, the gambling industry was rapidly growing, but a chance was underway. Increasingly, corporate capital was replacing the mob money that had funded casino property development in the past. Once Wall Street replaced organized crime as the financiers of economic growth in Las Vegas, any anxiety or apprehension about casino development in the city dissipated, and a growth mantra took shape that persisted until the global economic crisis struck Las Vegas around 2008. Throughout the 1970s and 1980s corporate entities like the Hilton Corporation, Caesars World, Harrah's, MGM, and Mirage Resorts blazed a trail for others to tap into corporate capital to develop casino resort properties.[5]

During the same period, the population exploded. All the construction projects created a massive demand for construction workers. Others were attracted to the city for its low tax rates, cheap housing, and other business opportunities.[6] Of course, this influx of residents created demands for more housing, which increased the demand for construction workers. By 1980, the city's metropolitan area had reached nearly 500,000 residents. Throughout the 1990s, Las Vegas was the fastest growing metropolitan area in the United States. In 1997, an average of 7,000 new residents moved to Clark County per month, many of whom were in search of better jobs, a better lifestyle, or a place to retire.[7]

State and local governments have a strong impact on the social, cultural, environmental, and political character of Las Vegas. Nevada has always been a state that finds bipartisan support for laissez-faire approaches to the public sector. Operating with "Western-style individualism," Nevadans have historically opposed "strong," formalized government structures, preferring instead a noninterventionist approach to governance.[8] This "frontier mentality" is reflected in the weak tax structure that is at once lauded for its attractiveness to companies, corporations, and investors but ignored as one of the main contributors to the region's shrinking public sector which has served as a major deterrent for businesses considering moving to Southern Nevada. Since this aversion to more interventionist approaches to public sector support has largely bipartisan support in Nevada, even when grassroots projects aimed at social, environmental, and social sustainability in the city are praised by local politicians, there is often little action taken at the structural level to facilitate, encourage, or incentivize this type of collective action.

There is widespread support for urban community gardening from city and county government by way of public praise for existing projects. When it comes to government support in terms of policies or programs to incentivize or promote more urban agriculture projects or to help communities with fewer resources bring gardens to their neighborhoods, the government is much less supportive. City Councilman Ricki Barlow who represented Ward

5, where VRCG is located, from 2007 to 2018, publicly voiced his support of community gardens in Las Vegas.[9]

Reflecting the region's Libertarian character, in Las Vegas, the fate of community gardens and other forms of urban agriculture is left to the market. There is an implicit assumption that gardens are great for the city, and that if they are meant to be in Las Vegas, the market will promote and support them; government intervention is not necessary and, in fact, antithetical to the promotion of "community." Clark County Commissioner Steve Sisolak was asked by the *Las Vegas Review-Journal* in 2011 about whether the city would consider a policy that would reduce or subsidize water rates for community gardens and he dismissed the idea saying, "I don't know how you justify that: You'll have people say, 'Well, my use is drinking' . . . Once you start picking and choosing, you're going to alienate a lot of people."[10] Sisolak's response acknowledges how many Nevadans might respond if the county intervened in the market-based provision of municipal water. Again, there is a deep history among the Vegas valley municipalities of refusal to pay for or contribute to projects, programs, or services that do not entirely or directly benefit them. This stance also characterizes many individual Southern Nevadans who demand the kind of Libertarian laissez-faire leadership mentioned above.

The words "urban garden," "urban farm," and "community garden" were completely absent from the Las Vegas municipal code until 2012. The only reason there was any change was due to a strong push to the Las Vegas City Council by Sarah, one of the garden's first long-term volunteers, a designer with experience in architecture and planning. She lobbied for months to get the city to introduce changes to city zoning that would allow for both smaller-scale urban gardens and larger-scale urban farms with animals and larger structures. Although VRCG participants consider it a community garden, if the proposed zoning variances were accepted it would be officially considered an urban farm. Since the council would have to introduce brand new municipal code, the goal was to convince the city to do so in a nuanced manner so that smaller neighborhood gardens would not be held to the same standards as larger, more complex projects, and vice versa.

After much back and forth between city officials, council members decided that in order to gain the support needed to pass zoning variances, the two-prong approach should be abandoned in pursuit of one initial code introduction. Sarah described the strategy of those council members supporting the change.

> And they thought that the city officials needed to get used to the idea of community gardens [first]. [. . .] And so we never really pursued getting a place like [VRCG] pushed through, and really it came down to nobody was messing with

> this garden here and we really didn't need to push getting any kind of zoning for this because nobody was bothering us. *But we're not legal.*

Sarah's comments illustrate a few points. First, in 2012, Las Vegas city officials had to "get used to the idea of community gardens." Sarah explained how some council members were concerned that without strict rules written into the code the garden would create blight in the neighborhood. Instead of advocating for comprehensive code changes and debating those challenging such changes, she claimed the council members who supported the zoning variances lied about the extent to which the variances for community gardens would cover VRCG. They then abandoned the variances that would *actually* cover VRCG and received widespread praise for successfully entering municipal code for gardening in the city.

The director of VRCG explained to me what she considered Councilman Ricki Barlow's most important contribution to the garden:

> Well he got the zone ordinance, and you know it didn't cost us anything. So he did see the importance of helping us to get legal so that we could do what we do in case when he's gone someone else tries to shut us down. (Roz)

As Roz points out, the passing of the zone variances would allow those who comply with the details of the code to not have to pay to get rezoned. What she does not acknowledge is that the passing of the zoning variance still does not cover VRCG. Therefore, if Councilman Barlow's future replacement is not a supporter of the garden, the property remains as vulnerable as it was before the variance was accepted.

Sarah's experience with the City Council debate illustrated the myopic nature of much local politics in Las Vegas by highlighting council members' position that since no one was coming after the garden now, there was no point dragging out a debate in order to create structural changes that would protect the garden into the future and make it possible for other projects like the garden to take off in other parts of the city where they may not be as well received. Furthermore, it calls attention to the vast differences between Las Vegas and a place like New York City where a great deal of research on community gardens is focused. These kinds of debates over zoning started decades ago in New York, and although the relationship between the city and the community gardeners has been fraught, in many ways, the conflicts have led to clarity for gardeners, sometimes resulting in periods of peace and autonomy, other times collective resistance. In Las Vegas, these debates are just starting, and the political will to figure out what urban agriculture zoning should look like in the city seems to be lacking.

There has been a great deal of buzz nationally about the educational, nutritional, and physical benefits of building gardens on school grounds and incorporating them into curriculum. This trend from the broader alternative food and food justice movements has gained more momentum in Las Vegas than others such as community gardens, community-supported agriculture, farmers markets, and local food co-ops. This trend, too, has been praised by local politicians, but ultimately left to free market.

In March 2013, a Las Vegas-based crowdfunding website GreenOurPlanet went live in its beta run and began connecting individual donors and corporate sponsors with Clark County School District schools whose students had requested the creating of a school-learning garden on their campus. According to their website, Green Our Planet has raised over $650,000 for Las Vegas valley school gardens, resulting in nearly 100 school gardens across the valley. Fund-raising campaign goals are typically met for a given school garden, either through direct corporate sponsorship (casinos find this a relatively cheap way to be associated with positive community projects), individual donations, or a combination of donation matching by a business. Once the money is secured, a company that normally helps Vegas area residents install residential fruit and vegetable gardens, called Garden Farms, steps in to help the school physically construct the garden and learn how to plant and take care of the crops before turning the gardening over to school faculty.

Owner of Garden Farms Bryan Vellinga appealed to the Western individualism so common in Nevada as he explained the virtue of this kind of partnership: "The beauty of it is no taxpayer dollars are used by it. It's all funded by the community."[11] Green Our Planet found a way to use similar approaches used by websites such as Indiegogo and Kickstarter to fund projects that ostensibly benefit the public sector without tapping into the tax dollars that make up the state budget. This is precisely the kind of market-based solutions local government publicly praises while largely avoiding the kinds of structural changes needed to support efforts like this.

THE NEIGHBORHOOD

Unlike the master-planned communities' characteristic of the Las Vegas valley and other Sunbelt cities, HWLV is a neighborhood that emerged as an ethnic enclave abruptly and unexpectedly upon Jim Crow era racism and segregation reaching Southern Nevada. HWLV is located within Las Vegas' Ward 5, but like many neighborhoods, it is best understood as a "natural area," or parts of the city that are an "unplanned, natural product of the city's growth."[12] A natural area is defined by its physical individuality and the cultural traits of its residents.

Unlike "administrative areas" like city wards that are arbitrarily carved out of cities, "railroad and industrial belts, park and boulevard systems, rivers and rises of land acting as barriers to movements of population tend to fix the boundaries of these natural areas."[13] Randy, a former garden participant who grew up in HWLV in the 1960s, described the contours of the neighborhood as, "three and a half square miles basically bordered by Bonanza, Martin Luther King Blvd, the railroad track, over to Carey Ave." The boundaries of the neighborhood do not conform to the political ones drawn by the city.

In addition to natural areas' distinct spatial character, Zorbaugh argues that variance in property values "sorts and sifts" urban dwellers into different natural areas. Additionally, he argues cultural factors such as race and ethnicity contribute to segregation in cities by

> [. . .] creating repulsions and attractions. From the mobile competing stream of the city's population each natural area of the city tends to collect the particular individuals predestined to it. These individuals, in turn, give to the area a peculiar character. As a result of this segregation, the natural areas of the city tend to become distinct cultural areas as well [. . .] each with its characteristic complex of institutions, customs, beliefs, standards of life, traditions, attitudes, sentiments, and interests.[14]

For over half a century, the neighborhood's residents have been predominantly low income and members of racial and ethnic minority groups.

History of Segregation and Neglect

Not all of Las Vegas was able to enjoy the economic growth that characterized much of the valley for the past seventy years. For instance, the neighborhood known as the Historic Black Westside or Historic West Las Vegas (HWLV) is a neighborhood that has for decades been overtly, then more covertly, denied services and left out of many of the social, political, and economic gains enjoyed by many other parts of the city. Although the racial structure of Jim Crow era racism took hold a little later in Las Vegas than it did in the Deep South, once it made it to the city it initiated a long, continuous history of segregation and deprivation for HWLV.

A century ago, there were roughly twenty black residents of Las Vegas. Although at the time some white residents were afraid that a racially integrated town might discourage whites from investing in the community, the low numbers of black residents kept whites from taking formal action to enforce residential segregation until the 1940s.[15] Blacks did face discrimination in Las Vegas, though. They were denied access to most brothels, they were denied jobs working on the dam, and before they were denied entry into nearly all casinos, they were kept out of some casinos on Boulder Highway that catered to dam workers. Most of the men building the dam had come from the South and brought with them demands of segregation.[16] Overall, though, the black population was not sizable enough to pose a threat to whites' "group position."[17] Blumer argues prejudice is, at least in part, a result of dominant racial groups fearing that a perceived subordinate racial group threatens their dominant racial position.[18]

When Basic Magnesium opened in 1942 in what is now Henderson, NV (at the time an unincorporated township in the Las Vegas valley), blacks from the South migrated to work in the plant. The black population jumped from 178 in 1940 to over 3,000 in 1943, and by 1955, there were over 15,000 black residents in Las Vegas.[19] As the black population grew in Las Vegas, and more whites from the South moved to the city, the relatively integrated city of the 1920s and 1930s adopted many of the Jim Crow policies already adopted by many U.S. cities. White residents began drafting "racial covenants" that legally barred black residents from purchasing land or homes in their neighborhoods,[20] the same federally encouraged practices widely used in Eastern and Midwestern cities that helped create the iconic black ghetto.[21]

These neighborhood-level actions to establish residential segregation were matched at the city level. Partially, not only a response to white residents' demands for segregation but also a concession to Southern tourists to Las Vegas' casinos, Las Vegas Mayor Ernie Cragin's administration began an effort to segregate black Las Vegans in the area now known as HWLV. Starting in 1943, black business owners were refused renewals of their business licenses unless that agreed to relocate their businesses to HWLV. Similarly, white landlords refused to rent to black residents in properties outside HWLV.[22]

At the turn of the twentieth century, Montana senator William Clark finalized plans for constructing a rail line to connect Salt Lake City and Los Angeles and purchased a large plot of land and its water rights to the east of the track. In anticipation of the new rail line, Las Vegas resident T.J. McWilliams purchased a plot of land West of the track. McWilliams developed his "Original Townsite," and it flourished as a camp for the town's service sector employees and a rest stop for those carrying goods and resources to and from nearby mining sites and boomtowns.[23] This changed in 1905 when Clark's rail line was complete and the "Clark Townsite" opened. Clark's rail company began paving roads, installing water lines to provide water to businesses and residents, and building schools, but only east of the tracks. Residents of the Westside began moving east, and shortly thereafter the McWilliams Townsite burned down, and the area remained neglected for decades.[24]

The location of the former McWilliams Townsite was the same area that black Las Vegas residents were pushed into by Mayor Cragin in the early 1940s. The living conditions of the area had not changed much in forty years. The homes were still little more than tarpaper shacks or tents, most of which had no electricity or running water, the roads were still not paved, there were no sidewalks, and no streetlights. As existing residents relocated to HWLV and thousands of recently arrived black residents moved into the neighborhood, the city did nothing to improve the living conditions, assuming that once the war ended and the jobs at Basic Magnesium and defense-related construction were no longer available, all the black residents would move back South.[25] The only options for many new black residents were to live in a tent, in their car, or build a small shack somewhere in HWLV.

The black residents of HWLV soon began forging a strong sense of community, and black-owned businesses thrived in the neighborhood. Jobs at Basic Magnesium paid relatively well, as did the "back of house" jobs at casinos that were available to black residents. As a result, the bars, casinos, retail stores, and restaurants in HWLV did well. In part, the development of strong community institutions was out of necessity since black residents were denied access to so much of the city. When desegregation finally gained traction in Las Vegas in 1971, it opened up residential possibilities and expanded access to public places for black residents, but it also inadvertently eroded the HWLV community institutions that were created during segregation.[26]

As the economic base of HWLV businesses dissipated, businesses closed, buildings were boarded up, and in many instances, nothing moved in to replace them. During the period of residential segregation, black residents from every rung of the social class ladder lived in HWLV. In the years since desegregation many of the professionals from the community have moved to different parts of the city leaving predominantly low-income residents in the neighborhood. In recent decades, the city has undertaken redevelopment projects throughout the valley, but the few that have been planned for HWLV have never managed to materialize. There is now a sense of distrust among HWLV residents of those currently in community leadership roles.[27] The collective experiences of oppression among HWLV residents during the era of segregation established a shared antagonist and a shared sense of purpose and urgency, resulting in a strong political base in the neighborhood.[28] Integration had the effect of fragmenting that base.

This mistrust was amplified when the city began with plans to widen I-15, which runs along the border of HWLV. Without telling residents, and violating their legal obligation, the City Council voted to close F Street, a major corridor in the neighborhood, and one that connected HWLV to the rest of downtown Las Vegas. In 2008, construction began on a massive dirt wall across the corridor where F Street passed under I-15. Again, the neighborhood residents were not notified. Eventually, the city replaced the dirt wall with a concrete one.[29] After years of protest and outcry, and organizing done primarily by black women in the neighborhood, the Nevada State Senate Majority Leader, Steven Horsford, who grew up in HWLV, was able to pass legislation that designated the neighborhood part of a historic preservation initiative, and initiated a process through which the city would pay for the removal of the F Street wall and contribute funding for economic development in the neighborhood.[30] F Street finally reopened in December 2014. The neighborhood's mistrust of the city was still high, but community leaders in the neighborhood felt optimistic after the victory getting the street reopened and the city's commitment to economic development in HWLV.

Struggles for Food Access

In the decades following integration, the struggle to attract businesses and redevelop HWLV has made it difficult for residents to access fresh food. In August 2004, the Vons grocery store that served the HWLV community since 1993 closed its doors citing "under performance." The national grocery chain was located on Owens Street between Martin Luther King Boulevard and H Street and was walkable for many low-income HWLV residents with little or no access to private transportation. At the time of its closing, there were two options for residents who wanted to shop at a supermarket, a Walmart and a different Vons, both of which were over four miles away and about an hour and a half walk from the Vons location. By bus, the trip to the alternative supermarkets took up to four hours roundtrip.[31] The Vons at Martin Luther King and Owens was not simply of material importance—providing neighborhood residents with fresh food—but also symbolic. The introduction of the grocery store in 1993 represented, at least symbolically, a new era of redevelopment reinvestment in a neighborhood that had been steadily deteriorating since racial integration efforts disrupted the relatively thriving West Las Vegas economy during Jim Crow era Vegas.[32]

After Vons closed in 2004, the ward's City Councilman Lawrence Weekly worked to find a chain to fill the void in the neighborhood. The city agreed to offer prospective stores up to $300,000 in tax rebates to open a store in HWLV.[33] Regarding the dearth of grocery stores in the neighborhood after Vons left, Weekly said, "It has been extremely frustrating for the residents to not have a grocery store in their neighborhood. Many of these residents are single mothers, seniors and others who do not have transportation."[34] After numerous failed attempts by Weekly and his successor Councilman Ricki Barlow, in 2009 the small West coast chain Buy Low opened in the previously abandoned Vons location. Although it allowed many without transportation to access groceries more conveniently, some black residents complained it catered mostly to Latino customers.[35] After a failed attempt in 2007, to bring a Food4Less to a location in HWLV, in 2012 a Dollar General store with a grocery section joined the Buy Low as the neighborhood's primary grocers.

The struggle to secure grocery stores in urban neighborhoods involves altering market dynamics in order to create demands grocery store chains deem worthy of their investments. Corporate decisions about where to locate new grocery stores are made largely based on "feasibility studies" to calculate the return on investment. Chains consider the size of the local market, local population, and income growth estimates, poverty rates and local access to assistance programs, local tax regimes, and access to public transportation, among other factors.[36] Some of these factors can be altered by local government, making neighborhoods more attractive to potential chains. As mentioned earlier, Las Vegas leaders tend to opt for market-based solutions. The struggle of HWLV to sustain access to quality fresh foods from grocery store chains cannot be explained as a failure in the marke,[37] Instead it should be understood as a failure in leadership and government.

Current Levels of Food Insecurity

Community food security requires more than just access and nutrition; it also involves affordability. Even when fresh, healthy foods are available, these options are often more expensive in low-income neighborhoods than in wealthier ones.[38] For the past five years, Three Square, a food bank serving all of Clark, Lincoln, Nye, and Esmeralda counties, has partnered with Feeding America, a nonprofit organization serving as a national network of food banks, to analyze hunger in Las Vegas. Based on data from 2013 that considered factors such as unemployment rates, median incomes, federal food assistance eligibility rates, and the average cost of a meal, the groups organized rates of food insecurity by zip codes in Las Vegas (see figure 2.1).

HWLV and VRCG are situated within the 89106 zip code, where 28.5 percent of residents meet the criteria for being food insecure, the highest rate in the city. This indicates that even though the neighborhood has recently secured two grocery stores, there are structural factors that continue to limit the ability of residents to purchase and consume healthy, fresh foods. This reality was, at least for a time, one of the motivating factors for starting a community garden in HWLV.[39] The landowner, garden director, some volunteers, and Councilman Barlow indicated at various times the void a community garden could fill for HWLV residents.

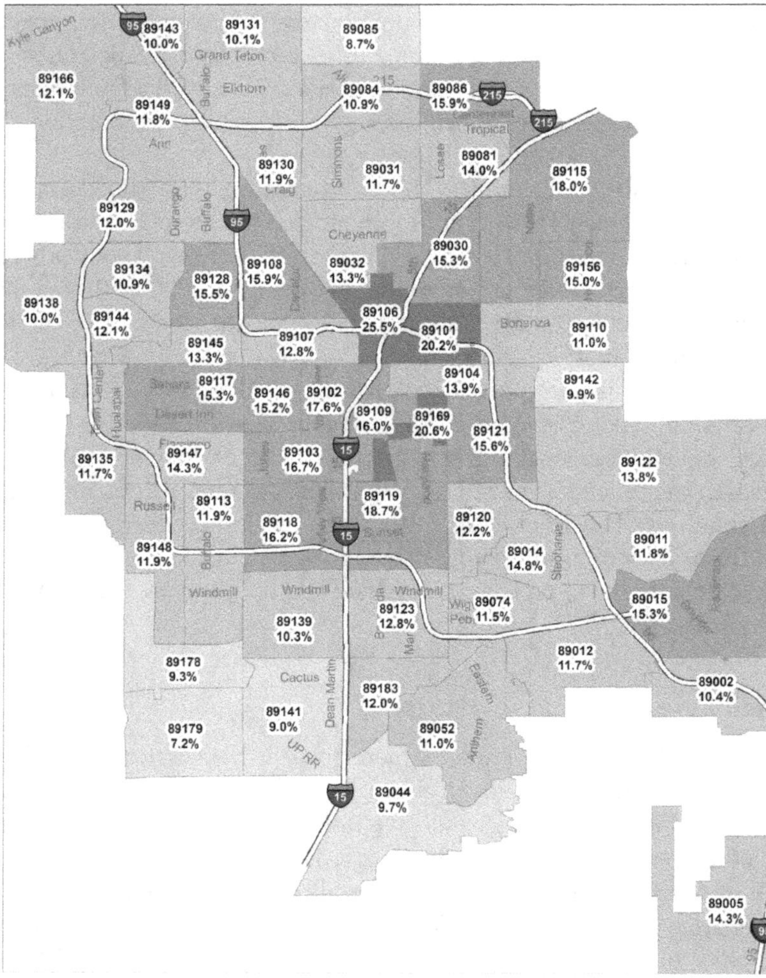

Figure 2.1 Food Insecurity in Las Vegas. The 89106 zip code was the most food insecure area in the Las Vegas Metro Area. In 2017, 25.5 percent of residents within the zip code were food insecure, down from 28.5 percent in 2013. *Source:* Threesquare and Feeding America's "Mapping the Meal Gap 2018" (2017 data)

THE GARDEN

Less than two miles from historic Fremont Street, on the edge of HWLV, lies a five-acre plot of land on North Tonopah Dr. sandwiched between an apartment complex on one side and a weekly–monthly hotel on the other. The property contains one duplex facing the road, a row of three small connected studio apartments and one detached studio apartment, a gravel parking lot, a small apartment used as a screen-printing business, and a large privacy fence bisecting the lot cutting off the majority of the property from the road. The fence is painted brightly with pictures of fruits and vegetables, a mural containing Aztec iconography related to agriculture and Latino farmers, and another mural of two dark-skinned women planting and watering seedlings. One of the units in the duplex is the office for VRCG. The property owner rents out the other unit and the studio apartment to various low-income residents.

VRCG sits behind a gate in the fence a few yards from the main office (see figure 2.2 for an architectural rendering of the property). As one enters the mulch-covered garden space, one passes by a paint shack on the left, and then a fenced in chain link "dog run" style cage that housed two goats before they were stolen (more on that later), and a small building on the right called the "resource room." The resource room contains a restroom, small kitchen, and storage for kitchen utensils used in demonstrations and events, as well as tables and chairs for classes and workshops. Next one passes two shade structures on the right covering homemade picnic tables and wrapped in misters for hot summer days. These structures are directly in front of a garage connected to the resource room where most of the gardening tools, seeds, the

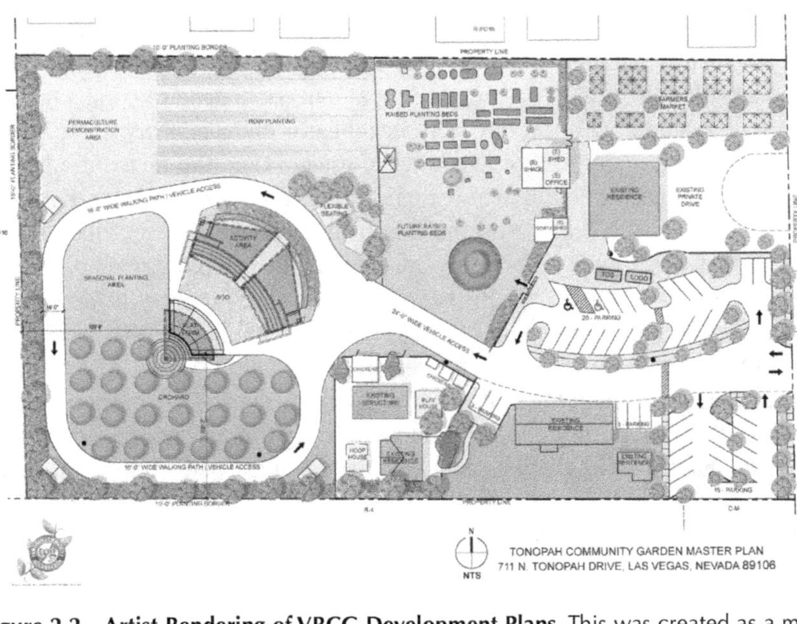

Figure 2.2 Artist Rendering of VRCG Development Plans. This was created as a master plan for the garden in 2011, and most, but not all of the plans, were realized. The plan to construct an amphitheater performance space by the orchard never materialized and neither did an on-site farmers market on the northeast corner of the property. *Source:* Artist rendering by Amy Finchem

stereo receiver, and a refrigerator are located. Next one enters an area covered in 36 10′×10′ raised bed planters. This is the "Adopt-a-plot" area, and nearly all of these planters are available for rent for $500 per year, including soil, seeds, water through timed drip-line irrigation, use of the garden's tools, and desert planting guide. The planters are made from cinder blocks, repurposed cement drainage piping, or plywood. There are also repurposed shopping carts and tires used to plant herbs and melons, a composting area comprised of an open-pit passive compost pile and a three-step active composting bin made from repurposed wooden pallets. Also in this area is a vermiculture bin (worm farm), which has gone through periods of both close attention and neglect throughout the garden's existence.

Near the back of the plots is a shed with a shaded work bench on the side that houses some of the more dangerous tools such as axes, an auger, chainsaws, and some organic fertilizers and pesticides, which are safe to use, but can be dangerous if children or animals get into them. This shed was built by a long-time volunteer and out-of-work construction worker, Jimmy. The roof is covered with two repurposed canvas advertisements for an amateur boxing match to keep out the rain. Directly outside the shed is a homemade brick and clay wood–fired oven.

Behind the shed and raised beds lie 21 in-ground planting rows like one would find at a typical farm. Throughout the year, participants at the garden plant seasonal vegetables and herbs in the rows, and the harvestable vegetables are for sale as a "You-pick" farmers market. Sales are conducted via the "honor system," and the director periodically updates a dry erase board with the produce ready for harvest and the prices. Like the raised gardening beds, these are outfitted with drip irrigation lines on a timer. Along the north side of the rows, there is also a makeshift-hoop house that provides shade for freshly planted seeds before they are transplanted into the rows or plots.

Beyond the rows, in the northwestern corner of the property, is a permaculture demonstration garden. This space is not maintained by VRCG participants, but instead by the Great Basin Permaculture Guild (GBP). The director of the garden agreed to let the group cultivate the corner of the property when she began the community garden. The goal of GBP is to create a space that highlights plant species native to Southern Nevada, food cultivation techniques that are more in-tune with the hydrological realities of the region (i.e., require less municipal water), and horticultural practices that come close to a "closed system" in which the by-products of one process become the inputs into the next, over time reducing the amount of external inputs necessary. In this space, GBP hosts demonstrations on these practices as well as native seed exchanges and an annual mesquite (a native tree) pancake breakfast.

Heading south along the western wall of the garden, one passes a large open space in the middle of the property with a fine red gravel track around it. This space was originally planned to be the site of an amphitheater for concerts, lectures, and other large public events hosted by the garden or other community groups. Garden participants have not managed to get the project off the ground, but it remains part of the lore of the garden. "Maybe someday . . . "

From the southwestern corner moving east along the southern wall, one encounters a fruit and nut orchard that contains the majority of the thirty-six fruit and nut trees on the property. These, too, are watered with drip irrigation lines on a timer. Continuing east, one encounters the Plexiglas Greenhouse in which most seedlings are housed in the colder months leading up to spring planting. Next to the greenhouse is a large mobile home where the property owner stores office furniture, and the garden director stores decorations for garden events. In front of the greenhouse and trailer are four chicken coops that contain nine egg-laying hens. In the middle of the coops is a large children's play structure with slides, ladders, swings, a climbing wall, and a playhouse.

The garden is never locked, so it is accessible any time of day on any day of the week. The garden website advertises hours of operation as Tuesday through Thursday and Saturday from 9:00 a.m. to 2:00 p.m.[40] These are hours when newcomers should expect the director or some other experienced volunteer to be present to field questions or explain what sorts of volunteer work needs done. Throughout the years, however, volunteer participation has ebbed and flowed resulting in stretches of time when visitors showed up during hours of operation and no one was at the garden, or at least no one who knew what was going on.

Roz

The director of VRCG, Rosalind "Roz" Brooks, grew up in Las Vegas not far from HWLV. She attended the University of Nevada, Las Vegas, and earned a bachelor's degree in business. After college, she managed a Kentucky Fried Chicken for about six months before having her first of two daughters in 1991 and taking on the role of stay at home mother for about ten years. In the late 1980s, while still at home with her daughters, Roz started a construction site cleanup company. By the early 2000s, Roz was going through a divorce with her first husband and needed additional income. She heard about a fast-track program through which individuals with bachelor's degrees could get certified to teach elementary school in approximately six months. Once she secured a teaching job, the school district paid for her to get a master's degree in education.

After a few years in the classroom, Roz felt compelled to get more active in the community, specifically with regard to the city's homeless population.

She got involved with a few organizations that served the homeless in North Las Vegas, while at the same time leading classes on health and weight loss at her church. Eventually, she quit her job to pursue opening a homeless shelter. In order to help fund the shelter, Roz started K&K Mobile Oil, an automobile oil changing company with the tagline, "We come to you for a change." The city ultimately refused to grant her the licensing she needed to open the shelter so she gave up on the idea.

Shortly thereafter, Roz ran into her cousin, Las Vegas native Frank Hawkins, at a business function. Hawkins played professional football for the Oakland Raiders from 1981 to 1987 and was the first African American to be elected to public office in Las Vegas history when he was voted in as city councilman in 1991. Hawkins currently serves as the president of the Las Vegas chapter of the National Association for the Advancement of Colored People and also directs the Community Development Programs Center of Nevada, a nonprofit organization that partners with the Department of Housing and Urban Development to build affordable housing in low-income neighborhoods in Las Vegas.

At the business function, the two talked about their shared interest in heath and nutrition and Roz's recent failed attempt at opening a homeless shelter. Hawkins told her about some land he owned and its potential for a community garden:

> When I met [with] Frank he mentioned that he had some land and people had been wanting to do a community garden on it and he said, "You can have it if you want it" and so I came out here and one thing led to another. Once I said yes it was on and crackin'. (Roz)

Roz decided to use the money she had been saving to open a homeless shelter to help start the garden. The garden would be located on land situated in HWLV. Hawkins previously used the property to store construction materials and discarded equipment from other endeavors, like calling card machines. In addition to donating the five acres for a community garden, Hawkins also paid the garden's water bill for the first year.

First, though, she needed to start a nonprofit organization.

Roz accepted Frank's proposal to start a community garden in HWLV in January of 2010. In March, she began filling out the paperwork necessary to establish her 501c3 nonprofit organization "Together We Can." Before the property could be cultivated, it needed to be cleared of all the remaining debris:

> This was his dumping ground for his construction material. So when he would get done with a site whatever was left he would just store it here . . . it was

fencing and rebar, and trailers, and . . . engines and just all kinds of stuff (laughing) . . . mounds and mounds of dirt. He wanted all that stuff, he was like, "Do not throw my stuff away!" You know, so we had all that we had to move it all over. So I got a U Haul and I got some day laborers. We just went from one side of the garden to the other, you know, moving it over [against the wall]." (Roz)

The procurement of day laborers to clean the site of debris to be ready for the creation of the community garden is extremely significant. It illustrates two things, first that the community of HWLV was not behind the project in the earliest days. If so, members would have been recruited to help in the site clean up. Second, it tells us something about the way Roz approached the task of creating a garden: as an obstacle to overcome. I will return to this second point below. I want to further highlight the significance of the dearth of community members helping in the clearing of debris.

Roz officially opened the garden to volunteers on March 13, 2010, and her nonprofit status was approved shortly thereafter. On the first day the garden was open, with the help of approximately eighty people, she planted pepper seedlings in the first fifteen raised bed plots and ten fruit trees around the lot. Since then the garden has been featured in all the major local newspapers and magazines, has appeared on every local television news station, and has generated media attention as far away as Washington D.C. Indeed, the garden has flourished and steadily grown, at least visually, but local community investment and support have barely picked up any momentum.

One of Roz's defining traits is her willingness to take leaps of faith. She described the importance of being able to take a chance, the average person's reluctance to do so, and her ability to set a goal in motion.

If you just make two steps the universe is gonna' make the next fifty. I mean we've seen that over and over at the garden. But you have to make a step. People will, they won't do that part. The first step is always the toughest, and to me that first step is just the easiest.

Others see this quality in Roz, too. From the start, Roz has had volunteers with professional expertise in areas relevant to the project. Sarah, an industrial designer, was one of Roz's first believers. Sarah and some of her peers in the local architecture and design community got to work designing the four acres using their knowledge of effective design elements, regulations, and zoning. Roz, however, had different plans.

I've always been a planner, you know? Even when we're out there talking with her we're *planning* of what this is gonna' be. And before we're even saying like OK here's where the beds are going she's out there building. Like, you know what I mean? That's just how she *is*. She's talking about those mini grants before we were even funded. (laughing) You know? (Sarah)

Importantly, in most of my interviews, people framed comments about Roz's reluctance to plan and wait for things to happen in a positive light. Roz herself seems to believe planning can be a way to avoid ever taking risks and even that some folks who plan too much for a project may not actually want to see their plan to fruition.

THE GARDENERS

There is no singular manner in which people have engaged with the garden. There is not an established or formal "volunteer" experience. Since the very beginning Roz has avoided clearly defining what is expected of those who want to get involved with the garden. Accordingly, there have been all sorts of different kinds of volunteers and participants throughout the garden's existence. I have identified six broad categories of participants: volunteer gardeners, office volunteers, plot renters, visitors, nonvolunteers, and core volunteers.

Volunteer gardeners are those who come to the garden ready to work and ask what needs to be done. These volunteers have helped build plots, move soil, water plants, paint, feed chickens, plant seeds and seedlings, harvest fruits and vegetables, and many other gardening tasks. These activities typically involve manual labor. This category includes individuals, groups of friends, families, school field trip groups, church groups, and many other types of groups who hold volunteer days as part of their membership. Despite a great deal of variation within this category in terms of how long and how frequently volunteer gardeners work, as well as how involved in decision-making about the work they are doing, what unites individuals in this category is that when they visit the garden they do so with the intention of working.

Office volunteers are those who help Roz with the nongardening tasks involved with running a community garden. Office volunteers have helped manage VRCG's social media and website, field e-mail inquiries, update the Adopt-A-Plot paperwork, conduct online research for future projects, plan events, and many other activities that help the garden function, but do not involve actual gardening. Sometimes office volunteers are recruited when an individual reaches out the to director to volunteer and she asks them if they have any experience with Microsoft Word, Excell, social media platforms, web design, graphic design, or other skills that would prove useful in helping to manage office tasks.

Plot renters are those who rent a raised bed plot but do not volunteer. They tend to visit the garden only to tend to their plot, and once they are done leave. Not all plots are rented by individuals, families, or friend groups. Companies and organizations have also rented plots, including Cox Communications, numerous churches, the Southern Nevada Health District, several local chapters of women's sororities, the Southern Nevada chapter of the Police Athletic League, and even the UNLV Sociology Department.

Visitors are the individuals and groups who come to the garden to find out "what it's all about." Often, when I encountered a visitor I said hi to them, and when they asked if I worked at the garden, I explained that I volunteer. I then asked if they had any questions, and almost without variation they asked some form of the request, "tell me/us what this is." Some of these folks had read articles in local news publications, others heard about it at church or through a coworker, still others saw the garden mentioned on Facebook or it had appeared in Google search results. The introductions to the garden are varied, but the curiosity that follows is strikingly similar.

Visitors are often not dressed for manual labor and are not interested in working at all. Instead, they are on a recon mission to find out if this place is for them; do they want to come back another time and perhaps contribute to the work underway? Some visitors turn into garden volunteers or office volunteers. Other times, they never come back. Sometimes, I would not see a visitor again for weeks or months and then they would appear again with a friend or group of friends saying, "See, *this* is the place I was telling you about." For some visitors, the garden serves as another stop on the Las Vegas tour, and for others, it is a place for periodic visits to relax, much like a park. There is quite a bit of variation within the category of visitors regarding how frequently individuals visit the garden and how their involvement changes, or does not. Over time, however, what unites them all is that they do not work.

Some people do work, but not as volunteers. Within this category are two major groups of people: individuals completing court-sanctioned community service and Latino day laborers. I call these garden participants *involuntary laborers*.[41] As a nonprofit organization, VRCG is an approved worksite for youth and adults completing court-ordered community service hours. The garden's website includes "community service opportunities" when enumerating the services it provides. When working with others throughout the garden property, it is common to ask others how long they have been coming to the garden of what brought them to the garden. This is typically when someone completing community service hours discloses their involuntary laborer status. Sometimes, they are comfortable talking about the process through which they were sentenced to community service and why they chose the garden. Some told me it seemed less boring than other options like working at Goodwill. Others explained they liked to be outside so this seemed like a decent way to complete their hours. One adult explained to me that he had been arrested and charged with a misdemeanor for growing marijuana, and he chose working at a community garden because he thought it was funny he could learn more about horticulture while fulfilling his state-mandated sentence.

One teenage boy who worked off his community service sentence a few days per week for over a month came to take ownership of the garden and the projects he worked on. He established positive relationships with other volunteers and the director, each of whom, myself included, saw a dramatic change in his attitude, maturity level, demeanor, and interpersonal skills. Long-term office and garden volunteer Sarah was moved by the change she witnessed in him:

> So when he left he had met that lady that owned a plot, and she really fell for the kid. I mean he was a total punk when he got here, but by the time he left he was this different kid, right? I remember by the time he left she was helping him to get his GED. He was going to get his GED, and his mom was going to get hers too, 'cause this lady was going to help them or something. And then he was going to go to culinary school, and I think he had said that they were going to start a garden there, and that he maybe was going to have some opportunity to help launch that. Gosh. Seeing that was really huge for me. He was shaking your hand hello. I remember he showed up to the wine [and cheese event] and he had just left, and he had come back for it and he was all dressed up and in a suit and he brought his grandmother and I didn't recognize him. It was so good to see him. I think it's the biggest transformation I've seen come from somebody's life being impacted from the garden.

Those completing community service hours are not working at the garden on a voluntary do not necessarily enjoy the work, but some like this teenager turned a court sentence into an enjoyable and ultimately transformative experience.

The other individuals whose participation is involuntary are the Latino day laborers periodically hired to complete manual labor tasks at the garden. In my drives through HWLV, I had noticed that just around the corner from the garden, a group of Latino day laborers gathered in front of an abandoned chicken restaurant on Bonanza Road across the street from a U-Haul rental station and a company that sells gravel, boulders, and decorative rocks for landscaping. In my first recorded interview with Roz, she mentioned that she paid some day laborers to help her employee with the oil change business clear off the lot in preparation for the garden, and although it struck me as an odd way to start a community garden, I was unaware that Roz relied on day labor to periodically give the garden a face-lift.

One day in May 2014, I arrived at the garden and noticed two older Latino men weeding in the rows and the fast pace at which they silently filled the wheelbarrows and discarded the weeds suggested to me they were not volunteers. I approached Roz and asked, "Who are those guys?" Roz was used to me asking about everyone at the garden, and she replied, "Oh I picked them up from Bonanza." She went on to say she couldn't keep up with the weeds. She said she hires laborers about once a month to do a whole garden clean up and then tries to maintain with volunteers in the interim.

Over the next few months, I noticed day laborers working a few more times, always doing the least "fun" tasks like clearing the rows in preparation for planting season, pulling weeds, or moving rocks and mulch. These are tasks that are hard, boring, and, again, and importantly, not fun. These are tasks I heard volunteer groups on numerous occasions complain about, suggesting they thought they would be doing things like planting seedlings, harvesting veggies, or watering plants. The tasks left to the laborers are not only less fun, but they are tasks that leave the garden looking well maintained. I will return to this issue and what it means in terms of the "community" aspect of the garden throughout the remainder of the book.

Finally, *core volunteers* are those individuals for whom the garden becomes a "third place." Third places are "the core settings of informal public life."[42] A diversity of types of public places can serve as third places for social actors so long as they "host the regular, voluntary, informal, and happily anticipated gatherings of individuals beyond the realms of work and home."[43] Importantly, locations that serve as third places for some people do not necessarily, or even likely, constitute a third place for all people who visit this place. Take, for example, the coffee shop or bar, both archetypal third places. For some patrons, like the "regulars" (whose presence are one of the defining traits of a third place), the location is a habitual hang out space where encounters and interactions are somewhat predictable, social differences are elided, and the mood is playful. For others, like the curious visitor, the location can be experienced as a confusing, intimidating, or dangerous place. The point is, there are certain characteristics that must be present, at least to a degree, for a setting to constitute a third place, but third places are also largely experiential phenomena. Core volunteers are those who tap into and latch onto the third place qualities of VRCG. It becomes a "home away from home for them." The mood is playful, they become a regular and can expect to see other regulars when they visit, even if they drop in unannounced. Oldenberg describes third places as "levelers," suggesting that in these spaces "conventional status counts for little," positions or ranks or hierarchies that are meaningful outside the space are abolished.[44] This feature is certainly present at VRCG among the core volunteers.

Core volunteers exhibit a degree of commitment to the garden that comes to resemble a sense of obligation and responsibility. Core volunteers are not just those who clock more volunteer hours than others. They are volunteers who take on additional responsibilities, tap into the institutional memory of the garden organization, make decisions about how tasks should be accomplished and hurdles overcome, and more generally, form the core of volunteer-based organization. Core garden volunteers are the people who can answer visitors' questions with certainty. They are those who can tell newcomers where to plant which seeds, where to find the chicken food, the combination to the shed's padlock, and when the drip irrigation waters different parts of the garden. Core office volunteers update annual budgets, server as garden liaisons to community groups, local government, and businesses, and prepare grant applications. Some core volunteers can do all the aforementioned tasks, while others have more narrowly focused roles.

Some folks have fallen squarely within one of these categories, but other participants' involvement has overflowed into numerous categories. These categories are what Weber calls "ideal types," so actual participants at the garden fit more or less within these constructed boxes. Throughout the remainder of the manuscript when I describe individuals involved with the garden, I will refer to these categories as touchstones to help orient the reader to the types of involvement different people have had in the garden.

THE IMPORTANCE OF TERROIR

The growing body of literature that looks for generalizable barriers to, benefits from, and best practices for community gardens throughout the United States and abroad has helped activists, scholars, and practitioners develop more complex understandings of community and urban gardening in general. What I do in this book, and the rationale for diving so deeply into the local context of this place-based research, is to illustrate how contextual factors, especially those of locations like Las Vegas that are so different than those well represented in the literature (e.g., Los Angeles, Oakland, Berkeley, Portland, OR, Chicago, and New York City), shape the manner in which community garden projects are produced, received, and struggle for survival.

The preceding mapped out the contextual backdrop, or *terroir*, from which VRCG has grown. Foodies and wine enthusiasts use the concept terroir to refer to the ways a place affects the taste of the food grown there. The reason wine from the Châteauneuf-du-Pape region of France, has unique flavor profiles, it is believed, is because of the very soil its grapes are grown in—its terroir. Something similar is not only true of community gardens in particular but also the practice of community in general. The soil or asphalt or concrete[45] where a community garden is planted matters.

NOTES

1. With her permission, I use Roz's actual name throughout the book since she is a public figure. All other interviewees are given pseudonyms. See Appendix I for a more detailed discussion on my research methodology.

2. This is the real name of the garden. It is a public place and is easily identifiable from a simple Internet search. I believe it is critical that, when possible, we use the real names of the places we study instead of creating pseudonyms because, as I argue throughout this book, the real particularities and context of a place impact how we interpret and understand them. For a broader discussion of my research methodology, see Appendix I.

3. An "acre-foot" is a measurement of water that is roughly equivalent to 325,851 gallons of water. The term literally refers to the volume of water it would take to cover an area of one acre at one-foot depth.

4. Eugene P. Moehring, *Resort City in the Sunbelt: Las Vegas, 1930–2000* (Reno: University of Nevada Press, 2000).

5. Hal Rothman, *Neon Metropolis: How Las Vegas Started the Twenty-First Century* (New York: Routledge, 2003).

6. Moehring, *Resort City in the Sunbelt.*

7. Mark Gottdiener et al., *Las Vegas: The Social Production of an All-American City* (Malden: MA: Wiley, 2000).

8. Gottdiener et al., *Las Vegas.*

9. Alan Choate, "Community Garden Faces Hitch in Its Water Supply," *Las Vegas Review-Journal*, June 27, 2011, http://www.reviewjournal.com/news/las-vegas /community-garden-faces-hitch-its-water-supply.

10. Choate, "Community Garden Faces Hitch in Its Water Supply."

11. Kristy Totten, "Garden Classrooms Project Sows Environmental Seeds," *Las Vegas Review-Journal*, July 12, 2013, http://www.reviewjournal.com/news/education /garden-classrooms-project-sows-environmental-seeds.

12. Harvey Zorbaugh, "The Natural Areas of the City," in *The Urban Community: Selected Papers from the Proceedings of the American Sociological Society, 1925*, ed. Ernest W. Burgess (Chicago: University of Chicago Press, 1926), 219–29, 222.

13. Zorbaugh, "The Natural Areas of the City," 222.

14. Zorbaugh, "The Natural Areas of the City," 223.

15. Moehring, *Resort City in the Sunbelt.*

16. Moehring, *Resort City in the Sunbelt.*

17. Herbert Blumer, "Race Prejudice as a Sense of Group Position," *The Pacific Sociological Review* 1, no. 1 (1958): 3–7.

18. Blumer, "Race Prejudice as a Sense of Group Position."

19. Moehring, *Resort City in the Sunbelt*; Rothman, *Neon Metropolis.*

20. Moehring, *Resort City in the Sunbelt.*

21. Douglas S. Massey and Nancy A. Denton, *American Apartheid: Segregation and the Making of the Underclass* (Cambridge, MA: Harvard University Press, 1993).

22. Moehring, *Resort City in the Sunbelt.*

23. Moehring, *Resort City in the Sunbelt.*

24. Rex J. Rowley, *Everyday Las Vegas: Local Life in a Tourist Town* (Reno: University of Nevada Press, 2014).

25. David Littlejohn, ed., *The Real Las Vegas: Life Beyond the Strip* (Oxford; New York: Oxford University Press, 1999).

26. Adrienne Packer, "Activists See Post-Integration Decay Continuing," *Las Vegas Review-Journal*, July 26, 2009, http://www.reviewjournal.com/news/activists-see-post-integration-decay-continuing.

27. Packer, "Activists See Post-Integration Decay Continuing."

28. Moehring, *Resort City in the Sunbelt.*

29. Robert McKee and Shannon M. Monnat, "Racialized Barriers and Social Action in West Las Vegas: The F Street Wall Controversy," *ASA Footnotes*, April 2011, https://www.asanet.org/sites/default/files/savvy/footnotes/apr11/vegas_0411.html.

30. McKee and Monnat, "Racialized Barriers and Social Action in West Las Vegas."

31. T. R. Witcher, "Hello? Can We Get a Store Here?," *Las Vegas Weekly*, June 30, 2005, http://lasvegasweekly.com/news/archive/2005/jun/30/hello-can-we-get-a-store-here/.

32. Jennifer Shubinski, "Councilman: Vons Closing 3 LV Stores," *Las Vegas Sun*, August 12, 2004, http://lasvegassun.com/news/2004/aug/12/councilman-vons-closing-3-lv-stores/.

33. Dan Kulin, "West LV Residents Rally for Grocery Store," *Las Vegas Sun*, July 8, 2005, http://lasvegassun.com/news/2005/jul/08/west-lv-residents-rally-for-grocery-store/#axzz2TsfGbkx6.

34. Kulin, "West LV Residents Rally for Grocery Store."

35. Packer, "Activists See Post-Integration Decay Continuing."

36. Alessandro Bonanno, "Food Deserts: Demand, Supply, and Economic Theory," *Choices: The Magazine of Food, Farm & Resource Issues* 27, no. 3 (2012): 1–4.

37. Bonanno, "Food Deserts."

38. Mark Winne, *Closing the Food Gap: Resetting the Table in the Land of Plenty* (Boston: Beacon Press, 2008).

39. Stephanie Tavares, "West Las Vegas Group Converts Vacant Lot of Land to Community Garden," *Las Vegas Sun*, March 22, 2010, http://lasvegassun.com/news/2010/mar/22/community-garden-sprouts/.

40. These hours change season-to-season—sometimes month-to-month—depending on volunteer availability and programming.

41. Claire Nettle argues that one of the defining characteristics of contemporary community gardens is that participation is primarily voluntary. For her and her research participants, the distinction is a matter of degree. If a garden is *mainly*

operated by people "contributing less-than-voluntary labor," it does not constitute a community garden—it is some other form of what Beth Pudup calls an "organized garden project." Participation at VRCG *is* primarily voluntary; however, the mix of voluntary and involuntary participation introduces obstacles to community building as I outline in later chapters. Mary Beth Pudup, "It Takes a Garden: Cultivating Citizen-Subjects in Organized Garden Projects," *Geoforum* 39, no. 3 (2008): 1228–40; Nettle, *Community Gardening*.

42. Ray Oldenburg, *The Great Good Place: Cafes, Coffee Shops, Bookstores, Bars, Hair Salons, and Other Hangouts at the Heart of a Community* (New York: Marlowe & Company, 1999).

43. Oldenburg, *The Great Good Place*.

44. Oldenburg, *The Great Good Place*.

45. Joy Santlofer, "Asphalt Terroir," in *Gastropolis: Food and New York City*, ed. Annie Hauck-Lawson, Jonathan Deutsch, and Michael Lomonaco (New York; Chichester: Columbia University Press, 2009), 174–94; John T. Lang, "Sound and the City," *Food, Culture & Society* 17, no. 4 (2014): 571–89.

Chapter 3

Miracle in the Mojave

Spiritual Place Narratives and Cultivating Community

The first time I visited Vegas Roots Community Garden (VRCG) I met a man and a woman tending to the raised beds as 1970s funk music played on a portable radio. The woman turned out to be Roz, who immediately put me to work dumping fresh soil into a new garden bed. Later, as I carried a fresh load of soil to the plot, I began talking to the man, Jimmy, about the soil. He told me about all of the different components contributing to the dark brown, fragrant soil. When I asked where the soil came from, he replied in an assured tone:

"We prayed for it."

"Oh yeah?" I asked.

"Mmm hmm," he replied.

That day Roz told me that I was, literally, a Godsend. She only recently began experimenting with planting directly in the ground in the garden rows and was worried she did not have enough human capital to prepare the rows, in addition to the raised beds and seedlings for the planting season. Roz said she had been praying for new volunteers and that my e-mail inquiring about how I could get involved was the answer to her prayers. At the time I had no idea how critical these initial exchanges would be for understanding the character of the garden as a "sacred" place, the goals of the garden organization, and the tactics employed to achieve them.

Literature on community gardens tend to focus on the perceived or potential outcomes of participation[1] or the impact of a community garden on the surrounding area.[2] When researchers focus on front-end matters involved in the creation and maintenance of gardens, they often emphasized practical matters such as the division of labor, resource mobilization, land acquisition, and the establishment of rules and regulations.[3] Using the case of VRCG, I extend the literature by exploring the ways in which cultural factors shape practical considerations in emergent place-based grassroots organizations. Focusing on the roles of religiosity and spirituality at the garden, I demonstrate how culturally mediated strategic choices come to shape and constrain potential outcomes of participation in such places.

Gardens and gardening have long shared an affinity with religion and spirituality. Indeed, the Christian origin story of humanity involves Adam and Eve's lives in and expulsion from the Garden of Eden in the Book of Genesis. The Bible and other religious texts rely heavily on plant cultivation metaphors to illustrate ideas about faith and divinity. Gardens have historically been places to practice devotion for a number of religions. Zen gardens, for example, have served as sites for Buddhist meditation for centuries. Gardens were also central to ritual and spiritual practices of the elite in Mesopotamia, Egypt, and Crete, exemplified by the Hanging Gardens of Babylon, one of the Seven Wonders of the World.[4] Currently, places like the Missouri Botanical Gardens in St. Louis feature Biblical Gardens where groundskeepers cultivate varieties of plants mentioned in the Bible.

Open spaces featuring "natural" elements have long been assumed to nourish individuals' mental and spiritual lives. The potential for recreation associated with time spent "out"[5] in nature has been also been increasingly attributed to pockets of "nature" in urban areas. The perceived importance of spiritually restorative green spaces in cities has led to the establishment of organizations like the TKF Foundation that, through their Open Spaces Sacred Places program, periodically funds the creation of tranquil, "natural" urban spaces intended to "provide sacred and spiritual experiences." The City Beautiful movement of the late nineteenth century included efforts by Protestant leaders and other "city beautiful evangelists" to promote the planting of trees, cleaning up and cultivating vacant lots, and construction of playgrounds and school gardens in an attempt to cleanse the moral character of urbanites and eradicate the "urban menace."[6]

SACRED PLACES AS URBAN CULTURE

This book is rooted in an urban culturalist perspective, which foregrounds the importance of "culture-place" relationships, and their centrality in helping residents both to experience their everyday urban existence as meaningful and to feel a connection to their cities and communities.[7] Community gardens are *places*, spaces made meaningful through interactions that have occurred in, on behalf of, or because of them. One of the ways urban space is transformed into a meaningful place is through the narratives people construct about urban places. Narratives are also central to the kinds of group culture that tend to emerge among interacting members of collective action groups.[8]

Narratives, Collective Action, and Place

The narratives we create and share about a place play a vital role in shaping our understanding of it and the things we do together in it. Broadly speaking, narratives are stories that "impose order on the flow of experience to make sense of events and actions."[9] A narrative is not a rational listing of chronological occurrences, but instead a meaningful contextualization of individual events within an overarching storyline. Narratives share a few common elements that aid in their explanatory, rather than simply enumerative, function. Events from the past are selected and organized into story form by way of "emplotment."[10] Organizing the past in this way allows for new experiences to be incorporated into the narrative by attributing meaning to them in terms of their relevance to the broader plotline.

The stories we tell about the "wheres" of our experiences and interactions are one way we transform undifferentiated urban space into meaningful places. As Barbara Johnstone points out, "stories do not simply describe worlds; stories also create worlds."[11] Place-based collective action, like the kind taken by VRCG participants, relies on narrative not only to make sense of their actions together but also to define the place, which in this case is both a product of their collective action and the setting in which it unfolds. As a place narrative takes shape, it can influence how people interpret or assign meaning to their experiences and interactions there.

The ways the past and present are constructed in a place narrative tend to narrow the "collective imagination"[12] or "horizons of possibility"[13] of those who use the place. Certain courses of action begin to seem obvious or inevitable, while others are rendered implausible or outright inconceivable. If we are to account for the human agency possessed by groups in our analysis of their collective action, we have to pay attention to the ways the group culture, or "idioculture,"[14] they create influences their decision-making. Narratives are central components of a group's shared cultural reservoir.[15]

Over time, multiple narratives tend to emerge in and about important places, making it necessary to examine the relationships between coexisting narratives. Sociologist Robin Clair identifies how multiple place narratives are not necessarily in conflict, but can function together as a "narrative thread"[16] that weaves through a community, binding it together. Clair describes communities as "narrative packages" in much the same way Gary Alan Fine identifies collective action groups as "bundles of narratives."[17]

One of the features of narrative that makes it such a powerful tool in the social construction of urban places is that narratives have a point. In their selectivity, organization, and tone they communicate not just facts but also themes.[18] This allows narratives to not only tell what happened but also explain or imply how and why it happened.[19] Place narratives are cultural creations rich with insights into how people become attached to some places and not others.

Sacredness and Place

Architects, urban planners, and social scientists from various fields describe people's attachments to and meanings for places as a "sense of place."[20] In his description of various instances of mid-century Bostonians acting toward urban space in noneconomic—or, nonrational—manners, sociologist Walter Firey identified the important role of sentiment and symbolism to urban experience.[21] He described how some urban areas take on a character of "sacredness," whereby the "profane" rules that govern the use of urban space elsewhere in the city do not seem to apply. The concept "sense of place" is derived from the Latin *genius loci*, which in its original form referenced the "guardian divinity" or "spirit" of a place.[22]

According to sociologist Émile Durkheim, humans always experience their own realities as bifurcated into two distinct realms: "the world of profane things on the one hand, the world of sacred things on the other."[23] He observed that "the energies at play in one are not merely different in their degree of intensity; they are different in kind."[24]

Sacralization, then, is the process through which objects, including places, are endowed with sacred properties.[25] The process of sacralizing an object or place involves an alteration, through interaction, of the way in which group members interpret its meaning. He argued "sacredness" is a quality individuals and groups endow upon things. Things, like the physical space upon which VRCG has grown, become sacred largely because they are treated as such.[26] The arbitrariness of sacred objects and places is summed up in Durkheim's claim that "anything at all can be sacred."[27] As groups bestow reverence upon a person, thing, or place, and come to attribute sacredness to it, it begins to exhibit those characteristics for them. Its sacredness is produced by the very veneration for it.

Sacred places in cities often, but not always,[28] involve elements of the natural environment.[29] In part, this has to do with shared cultural meanings of "urban" as dirty and unnatural built environments, and "nature" as wholesome and pure.[30] The perceived fundamental difference between these realms leads people to experience the presence of nature in cities as special, extraordinary, and sometimes even sacred. Even though gardens are part of the built environment since their existence reflects the domination of nature, they tend to be more closely associated with "wild" nature than the profane built environment of concrete and metal.

Examining the narratives people create and share about sacred places in cities helps illuminate the social and cultural processes through which urban sites come to signify and be experienced as sacred. As Christian Smith explains, "The sacred at the heart of any social order is always embedded in and arising from its collective narratives. It is the story that constructs the ideal. It is the narrative that defines the sacred.[31] Place narratives also, as we will see below, provide glimpses into the shared culture of groups like community gardeners that they rely upon to make sense of their work together and to guide future interactions.[32]

THE STORY OF A SACRED PLACE IN SIN CITY

Place narratives function to explain the meaning of the site and to render meaningful the actions of the people interacting in and on behalf of it. The dominant place narrative of VRCG revolves around the notion that the site is sacred and a conduit for divine assistance. Most linear narratives have a starting point, and VRCG's begins with a description of how a five-acre plot of urban space that was formerly, as Roz described it, a "dumping ground for . . . construction material" came to be interpreted and experienced as sacred urban space.

Transforming Vacant Space into "The Miracle Garden"

Origin stories and creation myths are crucial in the process of creating religious narratives about people, objects, and places, serving as touchstones in the collective "definition of the situation." The manner in which Roz explains how she came to quit her job as a schoolteacher and create a community garden established the tone for the garden narrative.

> I just really felt led to do something else. And so my husband you know we talked you know and he allowed me to quit and kind of follow where I thought God was leading, and he led me *here*. To be a *farmer* (Laughs).

With the genesis of the garden firmly rooted in an experience of divine guidance, the seeds of the place narrative were planted.

According to Janice, a long-time volunteer and friend of Roz, once the lot was cleared of debris, but before breaking ground, a small cadre of supporters gathered at the site and "walked it seven times and prayed on it." Janice said despite the original state of the property, she had faith in Roz.

> I believed it was her [Roz's] vision and whatever I could do to assist her with it that's what I did. So that night before the grand opening we walked this garden seven times and we was prayin'. Walkin' and prayin', walkin' and prayin', it was a few of us out here, walkin' and prayin'.

By walking the perimeter of the property seven times as they prayed over it, the otherwise profane urban space was consecrated. This interactive ritual served to modify the meanings associated with the garden space, and subsequent interactions there, for those involved.

Eventually, the general belief that Roz was led by God to start the garden, and that the site was sacred, developed into a more specific narrative. Many garden participants came to believe that VRCG was a conductor of metaphysical intervention and perceived the presence and influence of God, special "energy," or "the universe" on the property. Ever since the consecration ritual on the eve of the ribbon-cutting, phenomena at the garden were frequently interpreted through this lens. The accumulated stories of divine intervention formed a shared place narrative among garden participants rooted in the garden's origin story.

Roz shared the narrative of metaphysical intervention when she was interviewed in a *Las Vegas Review-Journal* article about the garden titled, "Faith makes fruits, vegetables sprout on five acres," and was quoted as saying, "This is the miracle garden . . . Anything I need, I just ask God, and, I kid you not, it just comes. When you have that going on, it's hard to quit."[33] Similarly, an article for the international web magazine *Grist* concluded with a quote from Roz, "I see the hand of God in every single thing we've done."[34] These exemplify the narrative of divine intervention that Roz constructed and employed regularly at the garden. They also illustrate how the spiritual place narrative was broadcast to a large audience, potentially shaping volunteers' and visitors' perceptions of the place before they arrived there, and attracting people with whom the narrative connected. For example, Peggy, a former plot renter and volunteer told me about how she first got involved with VRCG after reading the aforementioned *Review-Journal* article.

> There was an article in the newspaper [. . .] and it was an interview with Roz [. . .] I think the headline was something like, "Woman in faith believes garden can grow." [S]he talked a lot about her faith, and that resonated with me, and at the end of the story they published her phone number. So, I called her that day and introduced myself, and I just immediately connected with her and the vision.

Peggy emphasized her emotional connection to Roz's description of her faith-based approach to the project as the motivating factor for getting involved. The spiritual place narrative made sense to Peggy, and an affective bond was forged between the two women.

Sharing in the Sacredness—Merging Place (Narrative) and Community (Narrative)

Narratives operate transactionally, needing to be co-constructed between narrator and audience,[35] and if audiences are to identify with a narrative they need to be afforded opportunities to participate in the plot. As visitors and volunteers at VRCG encountered Roz's place narrative, they were invited to join an "interpretive community" and to see divine intervention at play at the garden. Narratives flourish when multiple accounts accumulate as plot points and support the overarching point of the narrative, and as more people adopted a shared "schemata of interpretation,"[36] regular volunteers began to describe scenarios in which material items, or a person with needed specialized knowledge showed up at the garden "just in time." Participants with these stories suggested there was no reasonable explanation outside of divine intervention.

Peggy: You hear, Roz herself, so many stories of "only God can." You know?
Tyler: What do you mean . . . ?
Peggy: "Only God can"? You've heard Roz talk about you know, like, she needed exactly seven posters for some pictures that were this odd size, 12 by 18 or something, and somebody showed up and said, "Oh I've got some frames in the trunk of my car." She popped her trunk, and there were exactly seven frames exactly twelve by whatever the crazy measurement was. Providence . . . The tractor that was donated just in time. Never later (Laughs). As God so often is, it's just in time. You know, the donations that come in just in time.

For Peggy, belief in both God and the garden created a space in which miracles become both possible and probable. Peggy made clear that, for her, the arrival of the frames was not a mere coincidence, but instead one example among many of providence.

As more participants witnessed events via the schemata of divine intervention, the interpretive community grew, and the spiritual place narrative filled out. Recounting divine intervention experiences at the garden, or participating in the spiritual place narrative, helps visitors and volunteers feel a sense of belonging via their shared experiences of place. One member of the since-disbanded advisory board, Casey, described her initial struggle to identify with the spiritual place narrative.

> Roz [was] always telling me that God would help and just like ask for help. And I was very stuck on that. I was like I don't understand that. That doesn't even make any sense. And then I started actually seeing miracles and shifting from fear to love and just seeing these miraculous things happen, and it really sparked

inside me. And now I consider myself a very spiritual person. And in that environment of the garden, I felt very safe to have those types of discussions.

For Casey, the spiritual place narrative at the garden was at first unintelligible. It didn't "make sense." Over time, though, subsequent interactions caused her to shift the manner through which she interpreted garden experiences, and she came to develop new meanings for the space, and she came to understand some of her own experiences as part of the place narrative. Casey not only came around to the narrative, she *became* spiritual. She also highlights the importance of "place" in her process of becoming spiritual. It was because she felt "safe" talking about divine intervention and miracles *at the garden* that she was able to develop her own spirituality.

The perceived sacredness of the garden extended to interactions that took place indoors on the garden property like advisory board meetings in the office. Casey described the meetings as being, "Almost like going to church." Casey drew a connection between her experience of the meetings, her broader understanding of the garden itself, and her own spirituality.

> I believe a lot in manifestation and divine guidance. And I remember Roz would say, "OK we need to pay this bill. I don't know how we're going to pay it, but I trust that it will happen. I prayed about it." And literally it would be like BOOM someone called her, and it was paid for. She's more coming from a religious point of view than I am. I come more from a spiritual point of view—but it was the first time I saw how you could co-create your reality . . . I saw miracles happen there. That's why I say it was like going to church. But it was like a magical, miracle (laughs) place. Every time I went to those meetings helped me believe more.

Casey's comments illustrate how the accumulation of stories about the miraculous things transpiring at the garden establish a coherent mythology that shapes subsequent experiences, like the interactions and emotional responses she had at meetings inside the garden's office.

Although the numerous accounts of "miracles" at the garden convey a history of divine interventions, they function to frame current actions, too. Hearing others' share the mythology of the garden and interpreting her own experiences at the garden through the lens of the garden's mythological narrative modified the meanings she associated with the garden and led her to "believe more" in both the garden itself and "manifestation" and "divine guidance." Over time those who believe in the sacredness of an object experience its impact on their lives as it helps them orient their broader spiritual beliefs and their beliefs in the object itself.[37]

The sanctity of sacred places also extends to the community of believers who inhabit it, who come to perceive the group itself and their actions as sacred.[38] As the garden narrative strengthened, there was a sense among group members that their actions were more powerful and potent when done for the garden. Sarah was also part of the advisory board and described feeling at one point that a different set of rules governed reality when it came to the garden.

> For a while Roz and I started joking that there's a whole lot of stuff we needed to start asking for because things just happened out here and there's really no way to explain it. And we started to understand that there was something really special about this project. Whether, that was whether you believe in God or you believe that putting really great energy into something produces really great outcomes.

Her comments do not simply reinforce the "only God can" mythology and the sense that not only the garden is a sacred place but also that garden volunteers were participating in its sacredness. Durkheim's assertion that the sacred totems of religious groups represented the best, ideal, and aspirational characteristics of the group itself, acknowledged that their perceived sacredness derives from individual group members' abilities to measure up to the traits of the group totem.[39] Sacred places help define community for those who participate and integrate them into the community.[40] For VRCG participants, the sharing of the spiritual narrative and situating their own experiences within that narrative was the vital cultural practice responsible for that integration.

The belief that miracles were happening at the garden, again, tended mostly to focus on the acquisition of needed materials, financial donations, or temporary volunteer expertise. In the heyday of the advisory board meetings, there was a palpable inertia mounting at the garden, and people wanted to be a part of it. The advisory board compiled a list of garden "needs," and they came pouring in. Like Sarah said, it really felt like anything we wanted we could simply ask for it and get it. Importantly, though, the seemingly miraculous works of God or "energy" at the garden were focused on *material* needs. According to community garden "how to" guides, though, these are some of the least important aspects of sustaining a community garden.[41] The garden also needed to establish organizational, ideological, and interpersonal norms,[42] and these elements only develop over time through repeated interaction. The garden narrative also shapes ideas about the garden's future, and since many believe God or some other supernatural presence guides the garden, there is a sense of relinquished responsibility for its longevity. For some, if garden participants asked for something and it did not come, it must not have been meant to be.

Narrative Compatibility—Secularism and Relative Sacredness

Not all participants employed the spiritual place narrative in the same manner. Even nonreligious participants referenced the garden narrative, albeit in more secular terms. The narrative drew strength from its ability to accommodate and integrate multiple interpretations that supported one another. Even when it was relayed in less spiritual terms, it took the form of describing the "specialness" of the garden relative to the rest of Las Vegas in way that closely mirrored Durkheim's famous distinction between the "sacred" and the "profane."[43] For many participants who talk about the garden in these terms, there is sense that the garden is a sacred space in the Durkheimian sense: "that which the profane must not touch and cannot touch with impunity . . . the two genera cannot be brought together and still maintain their separate natures."[44] Durkheim used the concept of "profane" to describe both evil elements of social reality and mundane aspects of everyday life. His dichotomy leaves little to no space for otherwise mundane spaces (urban garden) or activities (gardening) social actors anoint and treat as sacred. Las Vegas is recognized across the globe as a city associated with the profane. As well, the Las Vegas Strip is a reminder of the city's economic driver and profane cultural image that is visible from nearly any point in the valley. Importantly, though, only twenty-nine of the city's over seventy-five large-scale casinos are located on the Strip. The majority of the large casinos are located in neighborhoods. Additionally, there are "slot parlors" throughout the valley, and residents confront gaming while buying groceries, getting food or a drink at a bar, purchasing gas or snacks at convenience stores, and passing through McCarran Airport. What may seem exceptional if encountered in another city is mundane in Las Vegas.

Conversely, what may be seen as mundane in other cities, like a community garden or a quality public park, would be perceived as exceptional by some if encountered in Las Vegas. Wayne, a former plot renter and volunteer, described the garden as being extraordinary for Vegas.

> It's a pleasant place to be for me. It's very non-Vegas, and that's refreshing. [. . .] It's genuine. It's authentic. Instead of being artificial and manufactured. It's natural instead of plastic. I get pissed off. I don't know if you ever get out to Durango, and maybe they're doing this elsewhere, but in the median and along the edge now they're putting these artificial plants in, I don't know if they're plastic or metal. They're supposed to look like cactus. I hate it! Do that down on the Strip if that's what you want to do, but when you start doing that in my neighborhood, I get annoyed.

Wayne's designation of the garden as being "non-Vegas" is a common positive reaction among first-time visitors and veterans alike. Wayne's comments are echoed by Josh, who, although he identified as aspiritual, still described the garden as sacred, calling attention to its quality of "radical otherness" relative to the city's standard social and cultural fare.

> While I'm not religious, and I don't subscribe to the same spiritual beliefs the foundation of the garden is on, I felt the garden is a very spiritual place, so I accepted that while I may not agree on the religious principles that Roz or other folks may have, I still believe that this is a very spiritual and very special place in Las Vegas. It's different than anything else. (Josh)

Josh illustrates a common sentiment that by being "different than anything else" in Las Vegas, the garden takes on sacred qualities. Mary, another member of the advisory board, echoed the feeling, "I felt like I was part of something very special and rare . . . rare to Vegas." People are not simply identifying a quantitative lack of gardens in Las Vegas, but they are making a qualitative distinction between the character of VRCG and the dominant sociocultural landscape in the city. It is the radical otherness many attribute to the garden that imbues the place with qualities of the sacred. These non-believer gardeners are calling attention to the demarcative aspects of sacred urban spaces, or the power these places derive from their contrasts with their surroundings.[45]

The radical otherness of the garden is not just relative to Las Vegas but also to contemporary urban life. People describe the pleasure of being at the garden as "getting away" or "taking a break" from normal urban existence. These accounts highlight a common experience of liminality at the garden. Sacred sites are experienced as spaces existing between worlds, where time and perception operate differently,[46] and for some the garden is experienced as a site somewhere between the profane world of everyday urban life and the world of the metaphysical. For some others, the garden is experienced as existing between the mundane world of everyday urban life and the world of "wilderness," or "wild" nature that exists outside the city.[47] Jimmy, the head gardener at VRCG, went so far as to claim the garden protected him from the negative aspects of urban life.

> Yeah, we hear the cars go up and down the street, but they're so far off. It's a refuge for peace of mind. 'Cause, yeah, you can go crazy, you know [. . .] So, come here and just exhale. OK. I mean soon as I pull in the gate, I'm like 'I made it safe.' No police, no stray bullets, not wife yellin' in your ear about how no good you are . . . I can come in here and just be me. Yep. Don't pay the bills, but I got peace of mind."

Jimmy feels as though the garden provides him with both mental respite from the worries of everyday urban life and physical protection from the potential dangers of the city. Others describe is as their "sanctuary" and one volunteer, Joe, said "it's kind of easy to forget you're in the middle of Las Vegas. And I love that part." Taken together, these accounts describe the garden as a special place that radically differs from, and protects against, the dominant social and cultural forces in the city. Without using religious language, they ascribe a secular sacredness to the place, bolstering the claims of those who attribute the sacredness to divine intervention.

THE POWER AND PERIL OF STORIES

Religious myths and narratives contribute to both the space's stubborn persistence and its arrested development. Belief in a divine genesis of the garden helped Roz and other participants push through times of scarce resources. The perceived sacredness of the garden space and the garden narratives were buttressed by individuals' lived, embodied spiritual practices at the garden.

Faith served to construct perceptual blinders among participants, causing them to forego pursuing certain goals or lines of action. Roz illustrates this point.

> My faith has played such a big part because I don't stress about things, you know. I truly understand that if this is where God has me, God will provide, and when you really, really feel that way and really walk the walk and talk the talk then you don't stress, because if God closed it down tomorrow I would absolutely be OK with that.

Roz believes if the garden folds, it means God wants it to. Her belief in divine intervention is rooted in the assumption that God only provides when those seeking his help do not stress or act to meet the needs on their own. Such efforts, she claims, show a lack of faith and one should instead "say this is what I need and then forget about it." If God does not then provide, one should assume God does not think it is needed.

These beliefs result in multiple "blind spots" for her and the garden organization. First, relying on divine intervention for the provision of financial and physical capital has meant foregoing the development of consistent revenue sources or social capital through strong ties with retail or political organizations that could provide material resources needed at the garden. The garden lacks a committed, long-term core of volunteers, with frequent turnover among participants at all levels of commitment. Since Roz believes God will provide people when they are most needed, she does not focus much attention on building a sense of "community" or collective identity among garden volunteers and visitors.

A final issue related to the sense of community at the garden is not so much a blind spot, but a sore spot. The garden's immediate neighbors are predominantly low-income, black, and Latino residents, and the neighborhood has struggled for a decade to attract and retain businesses that provide quality fresh food. In the four years I have been involved at the garden, there have been few recruitment efforts directed at the neighborhood residents despite most participants agreeing that while the garden is intended to serve the entire Las Vegas valley, the main target demographic should be those who live nearest the garden space. This is partially due to the assumption that if God wants people to participate at the garden, he will see to it. The result has been the establishment of a broad, inclusive sense of who are part of the "community," and a nearly total absence of people from the immediate neighborhood.

Overreliance on metaphysical intervention prevented investment in infrastructure and outreach, and provided explanations and justifications for volunteer turnover and a failure to reach the local residents who live near the garden. The belief in the "only God can" narrative and the denial of responsibility contributes to the construction of these blind spots. The establishment of steady revenue-streams, coherent senses of community and collective identity, and connections with the underserved and underprivileged surrounding neighborhood requires the creation of a firm organizational foundation and infrastructure. The structure could take a variety of forms so long as it is consistent and stable. The establishment of such a structure is not prioritized at the garden, in part because of the mythology and narrative of divine guidance.

When people tell stories about miracles happening at the garden, they tend to focus on situations in which material goods, volunteers, and donations arrived at the garden "just in time." The "blind spots" are not issues simple enough to be dealt with through a donation or new volunteer. To be sure, money, supplies, and people are important, but more is needed. What is missing and cannot be provided by God or any other metaphysical being is the structure at the garden, and the relationships between those in the structure and those they seek to draw in. The organizational and interactional norms of any group are critical in determining the type of impact the collective can make. Narratives about a place arise out of interactions that take place in and about them. Over time these narratives can effectively shape the character of the place by altering the ways actions and events are interpreted, the ways in which people who gather there come to understand the place itself as well as the group, and perceptions about who the place is for, what it should accomplish, and how it should do so.

This research illuminates some of the ways in which cultural factors mediate the practical considerations involved in creating and maintaining grassroots community organizations. Individual and group cultural orientations, like spirituality, influence the goals perceived as desirable and attainable as well as the means believed to best pursue them. Accordingly, these same cultural factors play an important role in the impact such organizations have on participants and outsiders alike. Understanding the dynamic interplay between culture and place in shaping the character of grassroots collective action organizations is critical for analyzing the ways in which broad cultural trends like urban gardening are appropriated and developed in diverse local contexts.

NOTES

1. Rachel Kaplan, "Some Psychological Benefits of Gardening," *Environment and Behavior* 5, no. 2 (1973): 145–62; Tina M. Waliczekz et al., "Benefits of Community Gardening on Quality-of-Life Issues," *Journal of Environmental Horticulture* 14, no. 4 (1996): 194–98; Kimberly et al., "Leisure Spaces as Potential Sites for Interracial Interaction"; Diana C. Parry et al., "'Mary, Mary Quite Contrary, How Does Your Garden Grow?': Examining Gender Roles and Relations in Community Gardens," *Leisure Studies* 24, no. 2 (2005): 177–92; Kingsley and Townsend, "'Dig In' to Social Capital"; Mary Beth Pudup, "It Takes a Garden: Cultivating Citizen-Subjects in Organized Garden Projects," *Geoforum* 39, no. 3 (2008): 1228–40.

2. Brown and Jameton, "Public Health Implications"; Teig et al., "Collective Efficacy in Denver, Colorado"; Katherine Alaimo et al., "Community Gardening, Neighborhood Meetings, and Social Capital," *Journal of Community Psychology* 38, no. 4 (2010): 497–514.

3. Jeffrey Hou et al., *Greening Cities, Growing Communities* (Washington, DC: University of Washington Press, 2009); Bill McKelvey, "Community Gardening Toolkit: A Resource for Planning, Enhancing and Sustaining Your Community Gardening Project" (Columbia, MO: University of Missouri Extension, 2009), http://extension.missouri.edu/explorepdf/miscpubs/mp0906.pdf.

4. Jo Day, "Plants, Prayer, and Power: The Story of the First Mediterranean Gardens," in *Gardening: Philosophy for Everyone—Cultivating Wisdom*, ed. Dan O'Brien (Malden, MA: Wiley-Blackwell, 2010), 65–78.

5. Cronon, "The Trouble with Wilderness; or, Getting Back to the Wrong Nature."

6. Paul S. Boyer, *Urban Masses and Moral Order in America, 1820–1920* (Cambridge, MA: Harvard University Press, 1978); Laura J. Lawson, *City Bountiful: A Century of Community Gardening in America* (Berkeley: University of California Press, 2005).

7. Borer, "The Location of Culture."

8. Gary Alan Fine, "The Storied Group: Social Movements as 'Bundles of Narratives,'" in *Stories of Change: Narrative and Social Movements*, ed. Joseph E. Davis (Albany: State University of New York Press, 2002), 229–45.

9. Catherine K. Riessman, *Narrative Analysis* (Newbury Park, CA: SAGE Publications Inc, 1993).

10. David R. Maines, "Narrative's Moment and Sociology's Phenomena: Toward a Narrative Sociology," *The Sociological Quarterly* 34, no. 1 (1993): 17–38.

11. Barbara Johnstone, *Stories, Community, and Place: Narratives from Middle America* (Bloomington: Indiana University Press, 1990): 76.

12. Michael I. Borer, "From Collective Memory to Collective Imagination: Time, Place, and Urban Redevelopment," *Symbolic Interaction* 33, no. 1 (2010): 96–114.

13. Blee, *Democracy in the Making*.

14. Gary Alan Fine, "Small Groups and Culture Creation: The Idioculture of Little League Baseball Teams," *American Sociological Review* 44, no. 5 (1979): 733–45.

15. Fine, "The Storied Group."

16. Robin P. Clair, "Narratives in the Old Neighborhood: An Ethnographic Study of an Urban Neighborhood's Stories," *Qualitative Inquiry* 12, no. 6 (2006): 1244–61.

17. See, respectively, Clair, "Narratives in the Old Neighborhood."; Fine, "The Storied Group."

18. Maines, "Narrative's Moment and Sociology's Phenomena."

19. David R. Maines and Jeffrey C. Bridger, "Narratives, Community and Land Use Decisions," *Social Science Journal* 29, no. 4 (1992): 363.

20. Alex Kudryavtsev et al., "Sense of Place in Environmental Education," *Environmental Education Research* 18, no. 2 (2012): 229–50.

21. Walter Firey, "Sentiment and Symbolism as Ecological Variables," *American Sociological Review* 10, no. 2 (1945): 140–48.

22. John B. Jackson, *A Sense of Place, a Sense of Time* (New Haven, CT: Yale University Press, 1994).

23. Émile Durkheim, *The Elementary Forms of Religious Life* (New York: Oxford University Press, 1912), 160.

24. Durkheim, *The Elementary Forms of Religious Life*, 38.

25. Mircea Eliade, *The Sacred and the Profane: The Nature of Religion* (New York: Houghton Mifflin Harcourt, 1959); Douglas A. Marshall, "Temptation, Tradition, and Taboo: A Theory of Sacralization," *Sociological Theory* 28, no. 1 (2010): 64–90.

26. Durkheim's assumptions about how things come to be interpreted and treated as sacred depart from his largely functionalist foundation and enter into the realm of interpretivism. Ansom Shupe, "The Religious Institution," in *Handbook of Symbolic Interactionism*, ed. Larry T. Reynolds and Nancy J. Herman-Kinney (New York: Rowman Altamira, 2003), 625–36.

27. Durkheim, *The Elementary Forms of Religious Life*, 36.

28. Michael I. Borer, *Faithful to Fenway: Believing in Boston, Baseball, and America's Most Beloved Ballpark* (New York: NYU Press, 2008).

29. Eliade, *The Sacred and the Profane*; John P. Bartkowski and W. Scott Swearingen, "God Meets Gaia in Austin, Texas: A Case Study of Environmentalism as Implicit Religion," *Review of Religious Research* 38, no. 4 (1997): 308–24.

30. Scott Swearingen, "Sacred Space in the City: Community Totems and Political Contests," in *Research in Community Sociology*, ed. Dan A. Chekki, vol. 7 (Greenwich, CT: JAI Press, 1997), 141–69.

31. Christian Smith, *Moral, Believing Animals* (New York: Oxford University Press, 2003).

32. Fine, "The Storied Group."

33. Lynnette Curtis, "Faith Makes Fruits, Vegetables Sprout on Five Acres," *Las Vegas Review-Journal*, July 19, 2010, http://www.reviewjournal.com/news/faith-makes-fruits-vegetables-sprout-five-acres.

34. Greg Hanscom, "A Las Vegas Community Garden Beats the Odds," *Grist*, August 6, 2014, http://grist.org/cities/a-las-vegas-community-garden-beats-the-odds/.

35. Joseph E. Davis, ed., *Stories of Change: Narrative and Social Movements* (Albany, NY: SUNY Press, 2002).

36. Erving Goffman, *Frame Analysis: An Essay on the Organization of Experience* (New York: Harper & Row Publishers, 1974).

37. Borer, *Faithful to Fenway*.

38. Eliade, *The Sacred and the Profane*; Swearingen, "Sacred Space in the City: Community Totems and Political Contests."

39. Durkheim, *The Elementary Forms of Religious Life*.

40. Swearingen, "Sacred Space in the City: Community Totems and Political Contests."

41. Lawson, *City Bountiful*; McKelvey, "Community Gardening Toolkit."

42. Blee, *Democracy in the Making*.

43. Durkheim, *The Elementary Forms of Religious Life*.

44. Durkheim, *The Elementary Forms of Religious Life*, 39.

45. Bartkowski and Swearingen, "God Meets Gaia in Austin, Texas."

46. Bernadette Barton and Constance L. Hardesty, "Spirituality and Stripping: Exotic Dancers Narrate the Body Ekstasis," *Symbolic Interaction* 33, no. 2 (2010): 280–96.

47. Cronon, "The Trouble with Wilderness; or, Getting Back to the Wrong Nature."

Chapter 4

Talking the Talk

Performativity and the Cultural Production of a Community Garden

Like the variety of vegetables grown in them, community gardens take on many forms. I use the term "community garden" to describe many different kinds of spaces and projects in cities, towns, and rural areas. For example, the American Community Gardening Association identifies at least ten types of gardens, including neighborhood gardens, public gardens, public housing gardens, school gardens, senior gardens, therapeutic gardens, farms, youth empowerment gardens, giving gardens, and rooftop gardens.[1] Despite their diverse forms and functions, these varied projects can be understood as *cultural products* embedded in a shared cultural world. That is, community gardens are not created in cultural vacuums. They exist within a broader "green" cultural world,[2] and an urban gardening cultural trend with deep roots in American cities. Simply applying the label of "community garden" to a segment of urban space signals the creators' association or identification with the conventions of these broader cultural worlds. The name alone shapes audience and participant assumptions and expectations.

The literature on community development and placemaking often emphasizes the importance of participation among members of the target communities throughout the processes of developing and implementing plans for changes or development.[3] Community attachment, or detachment, to new or redeveloped urban places is a reflection of the process through which they are produced. This has tended to be the case with community gardens, with "handed-over" gardens often struggling to persist due to a lack of community involvement, and "buy-in," in the development stage.[4] Similarly, the literature on cultural production emphasizes how the coordinated actions of individuals in an organization and those within broader related social worlds shape the content of the cultural objects they produce.[5] Thus, a focus on how individuals' actions were coordinated to appropriate a broad cultural trend and develop it in a local context is a useful entry point to understanding the garden itself.

One of the most celebrated features of urban community gardens is their versatility. Residents and activists turn to gardens to address a variety of urban problems that affect diverse populations, using a wide array of methods.[6] Despite variations in mission, population, and approach, urban community gardens around the world look and feel very similar. In part, this similarity is due to a reliance on conventions and best practices that have emerged from the cultural worlds of the community gardening movement and sustainability discourse.

In this chapter, I illustrate how *over* reliance on communicative and linguistic conventions and *selective* reliance on organizational and administrative conventions contributed to a performance of place at the garden. I argue that the availability of cultural conventions makes possible the appearance of idealized standards that do not necessarily reflect quotidian realities, leaving the performance vulnerable to "disruption" and challenges from audience members and participants.

PERFORMATIVITY IN THE PRODUCTION OF PLACE

Places emerge from a combination of physical and social construction.[7] The former involves the manipulation of raw materials into relatively permanent structures and infrastructure, while the latter entails the creation of shared meaning in and about the place through interaction. Sometimes the individuals or groups responsible for the physical and social construction of a place are two distinct groups with, to borrow from the dramaturgical perspective, those in charge of the former consisting of "set designers" such as developers, architects, and property owners, and those in charge of the latter consisting of everyday people or "actors."[8] This distinction is not always so clear, however, as the case of place-based collective action around a community garden will illustrate. Sometimes actors are also involved in the physical production of a place, and the set designers are also actors in the place, and thus the social construction of the place is interwoven into its physical construction.

The physical construction of a place can also contribute to its social construction. *How* a place is built is meaningful. The sources of materials used, who is recruited to build the structures and the conditions under which they labor, and numerous other factors in the physical construction process shape the meaning of the place. The overlap between the physical and social construction of a place becomes apparent if one approaches places as *cultural products*.

It is useful to think about the construction of place as the *production of culture*. This requires consideration of the cultural contexts within which social actors make decisions about the physical construction of the place and the meanings they intend to express in and through the place. It also demands examination of the cultural contexts within which audiences and subsequent inhabitants of the place interpret and interact with and in it. Cultural products, like community gardens, are the outcome of cooperative activity between cultural producers[9] and support personnel, and this cooperation is mediated by *conventions* from broader cultural worlds. As Becker puts it,

> Conventions dictate what materials to use, [. . .] the abstractions to be used to convey particular ideas or experiences, [. . .] the form in which materials and abstractions will be combined, [. . .] suggest the appropriate dimensions of a [cultural product], [. . .] and regulate the relations between artists and audience, specifying the rights and obligations of both.[10]

Accordingly, cultural conventions are available for both the physical and the social construction of a place.

The production of culture perspective tends to highlight how relying on conventions helps cultural producers avoid extra costs (e.g., time, energy, material resources) that accompany going it alone or innovating through trial and error.[11] According to Becker, "We can understand any work as the product of a choice between conventional ease and success and unconventional trouble and lack of recognition."[12] For Becker, cultural producers have a choice of whether to adopt conventional means of physical and social construction of a product or take the more difficult, costly, unconventional route. However, conventions do not simply appear or exist; they emerge through prior interactions between individuals and groups attempting to accomplish a task.

Cultural conventions take shape once participants in a cultural trend arrive at a relative consensus about effective means to a collectively desired end. Conventions are a type of cultural norm or shared understandings of how to accomplish a task or communicate ideas and feelings. They emerge as "best practices" within cultural worlds. It is not so much that social actors have confirmed a given convention is objectively the most effective means to an end, rather, since cultural production requires the cooperation of many different moving parts, the conventional way of doing things is already understood by most participants in a cultural world. In other words, it is intelligible.

Often, it is the presence of conventions—whether material, linguistic, or organizational—that makes a cultural product intelligible to participants and audiences. Within the visual arts and music, certain mediums, techniques, scales, notations, and time signatures simplify the coordination of those producing the cultural product and render the product comprehensible to audiences. Accordingly, using conventions allows a producer to tap into existing networks of "support personnel" and "actors" in the cultural world and make it possible to "speak the same language" as audience members, because social actors often rely on conventions when interpreting cultural products. Conventions help render these products perceivable by appealing to the "group styles" of audiences.[13] Similarly, audiences "reading" a place look for conventional cues to help interpret what kind of place it is.

Intelligibility is simply an entry point through which audience members further interpret a cultural product. Over time participants in a cultural trend come to assume that conventions reflect certain orientations, express certain ideas, and promise certain experiences. Within a given cultural world, cultural producers engage in "boundary work" seeking to differentiate their products from, or align with, related products.[14] In this case, a group creating a community garden can employ certain cultural conventions in the physical construction of the site, the language they use to describe the garden project, and/or the manner in which the participants organize themselves and distribute labor and power in order to express that it *is* X type of garden, or that it *is not* Y type of garden. Even further specification is possible through the use of conventions by tweaking or customizing them for a local context.

Cultural producers tend to desire, and audience members and potential participants tend to expect, simultaneous integration into and differentiation from existing trends. Conventions attain relative stability among participants in a cultural trend, but maintain a degree of flexibility that allows them to be tweaked for a given iteration. Becker describes the rigid yet flexible character of cultural conventions within art worlds:

> Much of the content, symbolism and coloring of Italian Renaissance religious painting was conventionally given; *but a multitude of decisions remained for the artist*, so that even within those strict conventions different works could be produced (emphasis added).[15]

One aspect of cultural production highlighted here is that even when cultural producers employ conventions, social actors must still further specify the character of the new product. Employing conventions helps situate cultural products within broader cultural worlds and trends, but different production processes take place under specific conditions with diverse constraints, opportunities, and resources. In order to get the full benefit of a convention, producers must decide and articulate *how* they will use it.

When employed by cultural producers, conventions are able to elicit emotional responses among audience members and prospective participants. Through articulating what kind of cultural product, in this case what kind of place, is being or has been produced, cultural producers communicate the meaning of the place and impact how audiences *feel* about it. The adoption or rejection of a convention can evoke feelings of empathy, optimism, injustice, attraction, or repulsion, and these emotional responses impact subsequent actions among audience members. Meaningful places tell audiences stories about society by distinguishing good from evil thus defining morality and values. The manner in which cultural producers use conventions shapes what stories are told, what problems are diagnosed, and what solutions are perceived as plausible and just.

This chapter utilizes the concept of cultural conventions to examine the interplay between broad cultural trends like "community gardening" (i.e., culture at the macro level) and grassroots attempts at appropriating these trends in a localized context (i.e., the microfoundations of culture).[16] Examining whether or not, or how, social actors choose to adopt cultural conventions highlights the agentic elements of social action.

The choices producers of a place make regarding cultural conventions constitute *performances of place*. According to Goffman, the self is a collaborative social construction, composed of a strategically crafted image individuals perform for an audience and the interpretation and reaction the image elicits in the audience.[17] Whereas Goffman focused on interpersonal interactions between individuals and small groups, I am interested in the interactions between people *and* places and between people *in* places. Cultural producers of "place" engage in an ongoing performance similar to that of the performances of "self" individuals maintain every day.

In this chapter, I focus on dramaturgical behavior in a place-based organization. Dramaturgical organizational behavior refers not to individual dramatic behavior of individuals and teams within an organization, but instead to the collective presentation of the organization itself. More concretely, I am interested in the role of performativity in the cultural production of place in order to attract and retain volunteers, visitors, positive publicity, and funding. Importantly, the identity of the organization, or the definition of the situation, must be presented and supported not only by potential participants but also existing ones. That is, the organizational culture, or substance, should support the identity, or image, of the organization.

Cultural Conventions in Community Gardening

There is much more to creating a viable community garden than cultivating plants. As nearly every article or book on the topic attests, community gardens are as much about community as they are gardens. Accordingly, the cultural production of a community garden requires more than the construction of garden beds and hen houses, or planting, watering, and harvesting food.

Community gardening often requires creating and maintaining a budget, filing for non-profit status, writing grants, establishing a social media presence, creating rules and policies, recruiting and coordinating volunteers, planning events, developing community programs, soliciting donations, speaking in public, and more. Many people who start community gardens are passionate community members or city residents who lack expertise in some or even most of these areas. In the pre-Internet recent past, books like *Gardens for All* (1973), *Growing with Community Gardening* (1978), and *A Handbook of Community Gardening* (1982) provided those interested in starting a community garden with tips on both cultivation and practical advice on the aforementioned organizational and administrative concerns.[18] These books were published by urban garden collectives and individuals experienced in organizing community gardens seeking to help others avoid common pitfalls and potential oversights.

Currently, the Internet makes the sharing of community garden conventions much easier. A simple Google search can bring back thousands of YouTube videos on how to plant, grow, harvest and prepare fruits and vegetables, build raised bed plots, install irrigation systems, construct greenhouses and compost systems, and much more. Websites like Amazon.com reflect a booming market for the kinds of practical guides to community gardening that emerged in the 1970s and 1980s. Organizations like the American Community Gardening Association host websites with handbooks, fact sheets, tips, potential funding sources, and samples of documents like land-use agreements, garden plot registration forms, and garden rules. Many urban areas now have citywide organizations such as Gateway Greening in St. Louis, GreenThumb in New York City, or Seattle's P-Patch Program that provide physical assistance to community gardens in their vicinity as well as free online resources for community gardeners worldwide. University Cooperative Extensions often contain a Master Gardener division, which serves as an added resource for urban gardening expertise.[19] Additionally, many community gardens in the United States now have their own websites, Twitter and Instagram feeds, and Facebook pages, where they share information about their approaches to community gardening. All of these online resources provide the aspiring community gardener with models and best practices, or cultural conventions.

An individual interested in starting a community garden is able to adopt conventions that may be broadly understood as fitting into three categories: (1) material and spatial; (2) linguistic and communicative; and (3) organizational and administrative.

As the most basic of the three, *material and spatial conventions* refer to the built environment and appearance of the garden, strategies for and spatial organization of food production, and the "interactional potentials"[20] the place constructs for users. These conventions primarily cover the physical construction and maintenance of a community garden, but, as previously discussed, the decisions producers make regarding these issues are meaningful. For example, if a community garden organization decides to use recycled or repurposed materials for the raised bed, and to amend soils with compost created on site, the meanings it would associate with the garden would be different than if materials and soils were purchased from a home improvement store. Additionally, Kurtz has shown how decisions about fences, gates, and locks at community gardens alter the participants' experiences and conceptions of community.[21]

Linguistic and communicative conventions refer to the vocabulary generated by, or at least common to, the sociocultural worlds of the cultural product's intended audiences. Beyond the linguistic, communicative conventions also include the visual and audible stimuli associated with community gardens. These communicative conventions help producers convey information to others involved in the production process as well as audiences in a manner that is decipherable and resonant. In other words, it allows different producers and audiences within a shared cultural world to "speak the same language."

Linguistic conventions help transmit practical information about the cultural product, production process, and producer(s). Using them can also, purposefully or not, communicate broader abstract claims about the cultural product, production process, and producer(s). For example, if a community garden organization describes its growing practices as "organic," "GMO free," or "pesticide free," these buzzwords not only communicate practical matters about how the group cultivates food but also make moralistic points about genetic modification, chemical pesticides, those who engage in these practices, and, more broadly, those who eat produce grown under these conditions. Similarly, using words like "sustainable," "public," and even "community" to describe a community garden makes claims not just about the garden itself but also about other similar (and dissimilar) projects. When it comes to visual and acoustic communicative conventions, consider the various messages and claims a garden organization can convey through the murals, decorations, signage and other artwork it displays as well as any radio stations, recorded music, or live music played in the garden space. What one sees and hears, or does not, in a given community garden is meaningful and performative.

Organizational and administrative conventions are perhaps the most important of the three and refer to examples and best practices of leadership and participation, organizational structure and operations, decision-making processes, resource mobilization, and the garden's relationship to the community in which it is located and/or serves. These considerations are described in a popular community gardening tool kit produced by the University of Missouri Cooperative Extension:

> It is essential to understand that community gardening is about more than growing food, flowers and herbs. It's also about interpersonal relationships, group dynamics, planning and organizing, group decision-making . . . In short, community gardening is as much about "community" as it is "gardening."[22]

The centrality of this category of conventions lies in the critical impact that decisions about the division of labor, distribution of power, and interpersonal norms have on the ability to coordinate and cooperate to meet collective goals and ensure mutual gain for all stakeholders.

THE CULTURAL PRODUCTION OF VEGAS ROOTS COMMUNITY GARDEN

Since cultural conventions maintain a degree of flexibility in order to be adopted in various local contexts, I examine *how* Roz and core volunteers employ conventions. There are two dominant trends in the adoption of conventions at VRCG. The first refers to a propensity to adopt a cultural convention wholly, without serious consideration of how they apply in the local context. I refer to this practice as *overreliance* on conventions. The second is a tendency at times to adopt only a portion of a community garden cultural convention, leaving out elements that are critical to their effectiveness. I call this tendency *selective reliance* on conventions. Both tendencies have the capacity to undermine the effectiveness of a given cultural convention. The degree to which a convention must be adapted for a particular cultural product varies, but the process of customizing conventions to suit the production context is a critical one.

Overreliance

Total appropriation of a cultural convention is not necessarily problematic. The garden appears and semiotically reads like many of the most successful urban community gardens around the country. This is due in large part to decision-makers at the garden adopting conventions of urban community gardens in other U.S. cities. As Becker argued, this reliance on cultural conventions allowed the garden to build momentum much faster than if the founders had gone it alone and relied soon the process of trial and error.[23]

Roz described the first few months of the garden's existence as a period of confusion over what community gardening entails:

> I was coming every day. I was bringing in soil and seeds, and I did that for a couple months. It was very stressful, very hard, very tumultuous, because I didn't know what I was doing. A lot a lot of frustrating days.

Instead of relying on her own innovative spark, Roz turned to the Internet for assistance on urban gardening best practices. She described her daily routine during the spring of the garden's first season:

> I would just walk into this empty lot and just do whatever was on my to-do list based on what I learned the night before on YouTube or whatever.

Importantly, she was referring to material conventions, like the repurposing of automobile tires, shopping carts, cinder blocks, and concrete drainage channels (donated by a neighboring roofing company) into raised planting beds (see figure 4.1). Roz began by adopting conventions directly from YouTube videos, but quickly transitioned to adopting the conventional approach of bricolage, used by many community gardens and other community organizations.[24] This approach involves using what is available, piecing together tools and expertise from existing participants, and not waiting for the "right" materials or person for the job. Based on Roz's personality, it seems like this is the approach she would have taken anyway (see chapter 5). It just so happens to also be a common route for many under-resourced community gardens.

Randy, a retired school principal and core volunteer at VRCG, described another instance of exact replication of a material convention:

Figure 4.1 Raised Bed Plots at VRCG. Those in the background are currently in use, while the empty ones in the foreground were just built and waiting to be "adopted." *Source*: Photo by the author

> I read a very popular book [. . .] written by James Crocket called *The Victory Garden*. And with the victory garden I saw the diagram of making the compost bins, followed that model, and I've used it ever since then and that was in the mid-seventies [. . .] I decided to bring my equipment here and donate it and then start making compost out all the refuse that I knew we would have from such a massive project.

Randy's three-stage bins (see figure 4.2) facilitate "active" composting, which produces a nutrient-rich soil faster and in a more orderly manner than had participants experimented with other "passive" approaches.[25] These instances of convention replication are not necessarily problematic. They serve the time-saving function described by Becker and rarely require customization for contextual differences. One exception might be the adoption of *raised* planting beds and *raised* in-ground planting rows. Although these conventional approaches to growing are successful in yielding crops, some volunteers have posited that they are not sustainable, or at the very least ignore the hydrological realities of the region. Aaron, a garden volunteer and permaculture enthusiast, summed up this critique to me by suggesting that, given the enduring drought conditions in the American Southwest, in order to minimize the amount of water needed for plots it may have made more sense to implement *recessed* beds and rows to reduce run off and water waste.

As raised beds are much more common in community gardens nationwide, opting for recessed beds would have been more innovative and more labor- and time-intensive. When I asked Roz if she had heard about Aaron's idea for recessed beds, she responded by saying, "If he wants to do that it's on him." She was not interested in experimenting with unconventional approaches.

Linguistic/communicative conventions are more precarious, though, since communication and organization are widely cited in the practical materials produced by cooperative extensions and community gardening organizations as the factors that make and break gardening projects. Some linguistic/communicative conventions are institutionalized into social structures[26] like that of money granting organizations and media outlets. When a community garden applies for a grant, members are expected to frame their actions using words and phrases that are valued within the cultural worlds of the organizations. The same goes for the media. Journalists and reporters grant more credibility to certain institutional cultural schemas[27] or conventional ways of organizational presentation. Banet-Weiser acknowledges this reality suggesting that "while urban farming has been a source of food for varied communities for decades, in the current moment a specific version of urban farming is particularly brandable."[28] The idea that some community gardens can be more brandable than others illustrates why community garden organizations might adopt linguistic/communicative conventions in their performances of place.

Figure 4.2 Randy's Three-bin Composter. Organic, compostable materials are deposited in the right bin and watered. Once material begins to decompose, it is transferred to the center bin to continue breaking down and to free up space for new deposits in the left bin. Volunteers manually "turn" the compost to stir the contents, and water it frequently. Once it has broken down into rich, dark, matter, it is moved into the left bin, and ready to be added to soil in raised beds or the planting rows. *Source*: Photo by the author

Take, for example, the "About the Organization" crafted by Roz and other core volunteers for the garden's original website in the winter of 2010. Nearly every line in the descriptions of what the garden organization intended to address and accomplish, and why, was copied and pasted from the websites of other community garden organizations, the American Community Gardening Association, and the now-defunct Urban Garden Advocates organization.[29] To describe the mission and vision of Vegas

Roots, she used claims made by other gardens' websites about their own projects' aims and potential impacts of community gardening in general. These included goals like providing "job training and employment as apprentice gardeners for youth *from the local community,*" "a catalyst for *neighborhood and community development,*" and "increased access to healthy, high quality, fresh food *in the community, especially for low-income residents*" (emphasis added).

Over the next few years, the mission and vision statements underwent numerous revisions. First, Roz (or a volunteer) condensed the original description of the garden organization into a more concise version for grant applications and other official documents. The following is the original mission statement that was used for a few years until relatively recently:

> Our mission: Create a caring and productive community of youth and adults from diverse backgrounds who work together to build a sustainable food system and an eco-friendly environment. Specifically, Together We Can aims to
>
> - produce fresh food for residents of the city;
> - build young leaders from the surrounding community;
> - create neighborhood connectedness;
> - promote healthy bodies and minds.

The mission highlights a handful of issues of critical importance to community gardens across the country and utilizes language that suggests a degree of alignment with the alternative food and food justice movements. One of the most widely circulated definitions of food justice was crafted by the long-standing New York City organization Just Food: "Food Justice is communities exercising their right to grow, sell, and eat healthy food. Healthy food is fresh, nutritious, affordable, culturally appropriate, and grown locally with care for the well-being of the land, workers, and animals."[30] Furthermore, if one knows that the garden is located in the Historic Westside, the reference to "neighborhood connectedness" and "the surrounding community" suggests the garden was intended to be a "neighborhood garden" that is organized and managed by members of the surrounding community.[31]

Since the Historic Westside's persistent food insecurity had garnered media attention, local news outlets often framed the garden in the context of the neighborhood. The first news article about the garden—"West Las Vegas Group Converts Vacant Lot of Land to Community Garden"—explicitly describes the mission of the garden as attracting participants from HWLV and describes the neighborhood as a food desert.[32] Other articles also situate the garden within the neighborhood's history of food insecurity and feature quotes from Councilman Ricki Barlow describing the importance of the garden to HWLV residents.[33]

Some volunteers got involved because they assumed VRCG was, indeed, a neighborhood garden intended to serve as an antidote:

> I knew a little of the history of the community. How it's always been really segregated . . . understanding the history of the segregation and discrimination against the Westside, I really felt like there's no better community for the garden to be in. (Josh, core volunteer)

The linguistic conventions about community and neighborhood connectedness led some to make assumptions about some of the specific goals of VRCG. Moreover, Josh interpreted these conventions within the local context, eliciting an emotional response and evoking a sense of justice.

I asked Roz about the thought process that helped produce the mission statement and, by extension, purpose of the garden:

> I don't even know if that's, I, I think that's just what you *say* about a garden, you know? That's just like what *everybody's*, I mean what else are you going to say, like, "I'm just growing food to throw it away?" We're growing food and you know, to give it to somebody. (Roz)

Roz acknowledged that the mission is coded in a way that resonates with other community gardens, but not necessarily that she had employed conventions with specific cultural meanings within the social worlds of environmentalism and social justice activism. Understood as linguistic/communicative conventions, phrases like "sustainable food system," "eco-friendly environment," "neighborhood connectedness," and the emphasis on diversity are imbued with additional emotion and meaning,[34] and potentially carry promises about the garden, the organization, and its practices beyond or different than what the phrases indicate[35] outside the context of alternative food culture.

What I had not initially understood was that the mission statement itself (above) was borrowed, almost verbatim, from the Food Project, a community garden in Lincoln, MA, founded in 1991. This is not an inherently problematic approach, for instance, if Roz and her volunteers wanted to emulate the Food Project's approach to community and food cultivation. This was not the case, however, as Roz illustrated in her comment above. She viewed the mission statement as a formality, not a promise to current and potential participants. It was, instead, a sort of placeholder while Roz, essentially on her own, started building a garden.

As Roz started on the physical construction of VRCG and began its social construction via the linguistic/communicative conventions above, she did so without concretely, on her own or with other volunteers or stake holders, identifying what concepts like "community" meant at VRCG:

Roz: I hadn't thought of any programs or events or whatever, I was just gonna grow some food for the neighborhood.

Tyler: The neighborhood was the focus at that time?

Roz: Yeah probably. Yeah. Yeah it was more so umm just . . . well maybe not even the neighborhood, it was more so just growing some food. You know maybe I didn't even really know who was going to be the main user, but I had the lot and I started a community garden. The first thing was to grow some food, and once the food started growing then we started getting more traction, more attention, more media and then everything just kinda evolved from week to week, you know, and I never really you know I didn't put I didn't really have a plan. I didn't plan long term, it was just whatever happened from week to week. I didn't even try to make a plan, everything just evolved.

As this quote suggests, Roz may have unintentionally implied adherence to a material/spatial convention of a "neighborhood garden" focused on serving a food-insecure community.

The location, indeed, was selected in a manner similar to the selection of the mission statement. I asked Roz why, since she does not live in the Historic Westside, she decided to start the garden where it is:

> I have no idea because since this wasn't something I was aggressively looking to do and it just happened to be here and, you know, it kinda' just is what it is. I mean, you know, community gardens would work well anywhere, in any neighborhood.

When I followed up and asked about food access in the neighborhood, she dismissed it as an important issue:

> You know I don't think access is as much of a problem because everyone has a car. They can drive wherever they want. They *drive* wherever they want. [. . .] So short of not having a car, *then* you have a problem of access. But to say we have one grocery store in the neighborhood. There's another one two miles from here well hell you drive two miles and you might can't walk to it, you know, so really that's really not (laughs) you know, when you really get down to the truth of the matter it's *really* not an access . . . not in Las Vegas anyway, not in this community.

Not only did she perceive the garden's location in the Historic Westside as incidental, she also indicated that its purpose was not trying to address food insecurity in the neighborhood where it is located. Furthermore, by denying a heightened need for alternatives to market-driven food options in the neighborhood, she also ignored the racialized elements in food access and the demographics of the Historic Westside. However, this denial was visually contradicted in the garden.

As one pulls into the VRCG parking lot, one of the first things they see is a series of murals painted on a large wooden privacy fence that separates the parking lot from the garden. On this fence, there is a large section that features images of produce and words like "fresh," "vitamins," and "live life!" On another section, there is a painting of two brown-skinned women planting seedlings. The third section of the fence features a series of images displaying symbols of Latino heritage. There is a moon with the face of Coyolxauhqui, the Aztec goddess of the moon (see figure 4.3), on one end and a depiction of half of the face of Tonatiuh, one of the Aztec gods associated with the sun (see figure 4.4). Between these images are depictions of Latina women harvesting corn and a Latino laborer harvesting food in the desert on one side of a tree (see figure 4.5), and on the other a rendering of VRCG and a Latino farmer (see figure 4.6). The slogan "Sembrando Paz, Para Cosechar Justica," ("Sow Peace, to Harvest Justice") is painted above the tree (see figure 4.7). These images tell a story and establish a relationship between Latino spirituality, culture, and history, the Historic Westside, and VRCG.

There was a notable disjuncture between the mission statement, the visual imagery at the garden, and Roz's narrative about the goals of VRCG. This discrepancy is itself significant, and becomes more critical when one considers the reception of this cultural product. The rift between image and substance is even more pronounced when the producer believes the linguistic/communicative conventions are "just what you say," while the words imply an alignment with specific social justice and environmentalist norms and values. In this context, the performative production of place is vulnerable to loss of control over its impression.

For example, while Roz acknowledged that she and her volunteers are relying on conventions of community garden verbiage, she did not seem to realize she was making promises to her audience that may create expectations about the cultural product. Audiences, after all, tend toward faithfulness to others' performances, often believing the idealized claims of performers and interpreting them to be indicative of conditions that exceed the performance itself. Under these circumstances, Roz's ability to maintain expressive control was weakened as the "sign-accepting tendency" of the audience leaves them in a position to be misled by the image projected by the garden, compared to the substance of day-to-day operations.[36]

Selective Reliance

For the most part, throughout the year, the garden appears organized and productive because there are plants growing, mulch is spread nicely over

Figure 4.3 Fence Mural. A moon with a face resembling Coyolxauhqui, the Aztec goddess of the moon. *Source:* Photo by the author

the ground, few weeds are visible, and rotting food is removed relatively quickly from the rows. VRCG—or at least the "garden" portion of the community garden—is an undeniable material success. Someone visiting the garden may not see any volunteers or "community" members, but it sure *looks* great.

There are various discrepancies that may arise between appearances and actual activity in a performance. Goffman describes the genesis of one of them:

Figure 4.4 Fence Mural 2. A partial face depicting Tonatiuh, one of the Aztec deities associated with the sun, is on the left. To his right are two women harvesting corn, painted to appear Latina. *Source:* Photo by the author

Figure 4.5 Fence Mural 3. A Latino laborer depicted harvesting food in the desert. *Source:* Photo by the author

Figure 4.6 Fence Mural 4. A rendering of VRCG against the Las Vegas skyline. To the right, a Latino farmer. *Source:* Photo by the author

Figure 4.7 Fence Mural 5. "Sembrando Paz, Para Cosechar Justica" ("Sow Peace, to Harvest Justice") *Source*: Photo by the author

> If the activity of an individual is to embody several ideal standards, and if a good showing is to be made, it is likely then that some of these standards will be sustained in public by the private sacrifice of some of the others.[37]

For Goffman, when actors must sacrifice some of the ideal standards expected from a performance, the ones selected for omission are often those whose absence can be concealed for the sake of maintaining standards whose absence or misapplication cannot be concealed.

Telling me about her relationship to her parents' vegetable garden when she was a child, Peggy, a long-time core volunteer, made a telling observation about the community garden:

> I got to plant. Got to pick. Got to eat. Most of the hard work—sort of like here—I never pull any weeds here. I don't know, there must be garden angels that come by? Very few weeds here, but anyway. None of the really hard work. (Peggy, former plot owner and core volunteer)

This "hard work" of pulling weeds was, for the first few years, the purview of Jimmy, a carpenter who was unemployed due to the recession, who worked at the garden for sporadic and low pay. Eventually, Roz stopped paying him, and he quit coming as much. Additionally, since the garden was originally a gravel lot in the middle of a city in the Mojave Desert, there were not many weeds in its first few years. After years of showering the lot with water, it is increasingly more difficult to control weeds.

Even when Jimmy was coming regularly, maintaining a large community garden was too much work for just one consistent volunteer. This was largely why, as discussed in chapter 2, Roz began seeking the services of Latino day laborers. About once a month, Roz would pay day laborers to pull weeds, clear dead plants, and those gone to seed, move piles of mulch, and clear rocks and gravel from parts of the property. At least once at the end of the summer growing season and beginning of the fall planting season, these men were responsible for enabling fall planting to proceed on schedule. On some days only Roz, the laborers, and I would be present. On others, there would be volunteers and people working in the permaculture garden. On these days, the volunteers were put to work planting in the rows and other more fun tasks while the laborers did the "hard work" of weeding and maintenance. There was almost no communication between the two groups. Although these men may have come from the surrounding neighborhood, the terms of their presence at the garden did not generate a sense of community connectedness.

The perceived need to pay day laborers to conduct basic garden maintenance both contributed to and resulted from the lack of a firm foundation of committed community volunteers. This is in large part due to the selective reliance upon one of the most common cultural conventions in contemporary community gardening: requiring the space to be *user created.* In the 1970s, amid a community garden resurgence in American cities due largely to a rejection of chemical pesticides and a growing environmentalist movement, many cities began investing money in the creation of community gardens for residents. Some were user-created, but others were built by the city with the assumption that community members would take over control once completed. Many of these "handed-over" gardens were quickly abandoned, in part because the communities they were intended to serve were left out of the development phase. Before long, organizations interested in funding gardens began insisting that the intended garden participants "reach a certain level of internal organization and be able to define its needs before proceeding with the garden."[38] Today, one can see this insight articulated by the American Community Garden Association in one of their five core beliefs:

> In order for a garden to be sustainable as a true community resource, it must grow from local conditions and reflect the strengths, needs and desires of the local community.[39]

This sentiment is summed up by organizations like the National Recreation and Park Association's *Grow Your Park Initiative* who urge interested parties to "build your garden community before you build your community garden."[40] In her eagerness to start physically constructing the garden, Roz refrained from organizing the intended, or even interested, users of the space.

Roz told me when she received approval to start working on the garden her first move was to clear the debris off the property.

> I got a U Haul and I got some day laborers. We just went from one side of the garden to the other, you know, moving it.

This first step was a critical one not only by starting the project before establishing a committed community but also by establishing a precedent that a reliance on day laborers for the functioning of the garden is not antithetical to "community" gardening. Creating a sense of community ownership of a project like a garden is often more dependent upon participation in the process of its development than access to the final product,[41] and preparing and developing the garden site is a vital step in the process of starting a community garden.[42]

Roz quickly established a small core of interested people. She met Sarah, a local designer, at a grant writing workshop who developed an interest in the project. Roz also enlisted a few of her friends and a handful of people who had read about the garden in the newspaper or heard about it on local radio. With their help, Roz was able to organize a groundbreaking event attended by approximately eighty volunteers. Over the next few months, Roz worked at the garden in almost total isolation, with nearly no volunteer help. The groundbreaking turnout had been greatly influenced by Disney's "Give a Day, Get a Day" volunteer promotion. Volunteers were neither neighbors nor even necessarily interested in committing to the project.

Sarah eventually brought a few of her friends in the design community on board to help create a site plan for the garden. Their training prompted them to propose a series of design charrettes with children, families, and seniors from the surrounding neighborhood. As she explains:

> We had gotten some kids that were living in those [neighboring] apartments together [. . .] And then asked them what they would like to see up here. [. . .] [A]nd I think the big consensus from the kids was they just wanted to play with bugs, and they wanted big piles of dirt to play in and it was very, it was not what vegetables they wanted at all. Really, they just wanted bugs and dirt to play with. And then I think Sherry did one more of like we were trying to get people in the local area here like senior citizens. I think Roz reached out to the senior center close by and got some out here [. . .] and kind of talked about different things they would want to see. (Sarah, core volunteer)

All of these community input efforts took place in the spring of 2010 and were never repeated or followed up. The primary outcome was the development of a professional master plan for the property.

Despite the unpredictable rates of volunteer turnover, throughout the first few years, Roz and other core volunteers were opposed to allowing local "experts" in the areas of gardening and nonprofits to get too involved in VRCG. Expertise was welcomed, but the experts themselves were perceived as outsiders and not welcome to participate in the decision-making process. This was an attempt at keeping the garden user-created. The power to make decisions at the garden was held close.

In the fall of 2010 and spring of 2011, the garden picked up speed as many people from outside the immediate neighborhood heard about the garden on the radio or via one of the local newspapers, and were getting involved as volunteers and/or plot owners. Almost immediately, however, it was apparent there was a volunteer retention problem and even the most committed volunteers, nearly universally from outside the immediate neighborhood, would get burnt out or too busy to continue with the project. From the fall of 2012 to the present, participation has declined rapidly. There are many vacant raised bed plots and many beautiful Saturday mornings pass with no volunteers or visitors.

It was under these conditions of an inconsistent volunteer base with relatively shallow roots that Roz felt compelled to hire day laborers to do the "hard work." Under optimal conditions, this is work that committed community volunteers would feel compelled to do. By relying on involuntary laborers at the outset of the physical construction of the garden, VRCG was able to start taking shape before a "gardening community" was established. Subsequent rejection of expert involvement in favor of relying on inconsistent volunteer help was decided in the name of keeping the garden user created. This selective reliance on the user-created convention led to the perception that in order to keep the garden *appearance* in order, the user-created standards should loosen and involuntary laborers could help pick up the slack. Ideal standards of garden appearance could be controlled more than ideal standards of community. Outsider involvement from involuntary laborers could be controlled more easily than that of experts.

PERFORMANCE DISRUPTIONS AND ACTUAL SOCIAL IDENTITY

Similar to individuals presenting a "self," cultural producers always have an intended audience in mind. They incorporate the perceived values, beliefs, and norms of this audience during the production (and reproduction) process and into the product. Accordingly, cultural products are "designed as much as possible to reflect [audiences'] tastes, interests, and attitudes."[43] One way this is done is through the reliance on cultural conventions, which must be carefully employed if the producer is to maintain "expressive control" over the meanings assigned to the product, production process, and producer. Since conventions typically articulate the cultural worlds whence they originate, audiences, with their "sign-accepting tendency"[44] may read into the product, process, and producer meanings that were not intended, setting up expectations that may not be met. In this way, relying on conventions can lead to "misrepresentation," or the impression of "the presence of something that is really not there."[45]

Cultural production is also vulnerable to the tendency of idealization. Just as Goffman pointed out for individuals, organizations also have a "tendency for performers to offer their observers an impression that is idealized in several ways."[46] This tendency may in fact be amplified in cultural production as producers are already trying to appeal to the norms, values and beliefs of their audiences. Under these conditions, cultural producers are vulnerable to discrepancies between the image projected to audiences and the quotidian substance of the cultural object, production process, and/or producer.

Conceiving cultural production as performative exposes some vulnerabilities to the reliance on cultural conventions. They do not simply save time and other scarce resources in the production process, but they also introduce important variables to the process that must be carefully managed to avoid what Goffman calls "performance disruptions," or instances when the discrepancy between image and substance, or front stage and backstage, is made visible.[47]

Performance disruptions occur at VRCG when audience members become aware of the discrepancies between the image Roz projected, intentionally or not, the garden organization, and the reality of daily garden operations. Impressions are "delicate, fragile things that can be shattered by very minor mishaps."[48] Because audiences tend to err on the side of faithfulness to the claims made by performers, performance disruptions can lead to a loss of reputation and credibility. However small, a disruption can also lead audiences to assume the performance is faulty in other, perhaps larger capacities.[49]

Wayne, a garden volunteer and plot owner, told me he got involved largely because he lived in a condo with very little space to grow food, but that he was also attracted by the garden's mission to provide food for the hungry people in the neighborhood. Over time, he grew frustrated with how much produce cost and the $500 price for renting a raised bed plot for a year. He said he did not understand, "Roz and her Mercedes parked out front . . . I don't get it." He went on:

Wayne: It's like please, I'm paying a hundred twenty-five bucks for a quarter here . . . . how much of that is paying for that Mercedes, you know? [. . .] I wonder what the water bill is a month.

Tyler: I think it's around $8,000 a year.

Wayne: How many plots are there?

Tyler: About thirty.

Wayne: Alright, so what kind of money does that bring in? Say thirty. Fifteen thousand. Alright so half that goes for water. What's the other half doing?

Wayne was frustrated that his initial impression of the garden was wrong and that poor residents in the neighborhood could not afford to participate. This led him to question the larger integrity of the community garden organization. Frustrated and deterred by the cost of participation, he decided to leave VRCG and joined a similar, cheaper garden affiliated with a Mormon church.

Recruited as a member of the "working board"—a group of core volunteers who help the garden create organizational infrastructure, plan events, and raise funds—Lisa also grew frustrated that the group was not as democratic as implied:

> [The] organization's focus, mission, name, priorities, etc., seem to change rapidly according to Roz's whim. Accordingly, someone's donation may go to something completely different than they thought at the time of donation. There is a lack of transparency and accountability with respect to finances at the garden. The more time I've spent at the garden and come to understand it better I've became *less* inclined to donate.

Lisa was frustrated because she believed that the working board would be democratic and that the garden's direction and focus would be user-generated. Over time she began to feel like it was not democratic in practice and started to doubt other elements about the garden. She eventually left the garden and started volunteering with a school garden.

Josh described getting involved to serve the residents of the Historic Westside only to later realize the garden was more of a platform to promote a vegan lifestyle:

> I felt at that time Roz was really kind of shifting from the educational and kind of serving the Westside community focus to we need to teach people how to eat healthy and clean and focus on individual health. [. . .] I really kind of felt like OK, while that is something I feel is really important that's not why I got involved in the garden, that's not my passion, that's not something I want to dedicate a lot of my time to because I feel like that's a me, personal decision. I don't really want to force that on other people. I want to educate other people about the benefits of growing their own food and fresh fruits and vegetables, but I don't want to say this is how you must eat. So, and I kind of felt that that's kind of where we were going and that's not a direction that personally *I* could subscribe to.

Although he began to disengage initially for unrelated issues, Josh decided to completely stop because of the gap between his initial impression of the garden and the very different reality he encountered later. He also left the garden.

GROWING WITH WEAK ROOTS

Community gardening cultural conventions have undoubtedly contributed to the persistence of Vegas Roots. As this case illustrates, however, the decision of whether or not to adopt cultural conventions is more complicated than simply opting for "conventional ease" over "unconventional trouble" as Becker suggested.[50] It is true that adopting conventions from relevant social worlds helps cultural producers save important resources such as time, energy, materials, and money, which are typically lost when one attempts to go it alone or innovate.[51] Trial and error is typically a resource-intensive approach. Conventions do not simply constrain innovation as Becker asserted but also enable one to bypass critical stages in the production process that are crucial for establishing a foundation for sustained production.

Conventions can also contribute to the misrepresentation of the cultural product and the production process itself, creating a rift between the impression presented and the day-to-day reality. This chapter illustrates how disjunctures emerged from linguistic and communicative conventions and organizational and administrative conventions. In this case, the adoption of cultural conventions created complications in the consistency between the stated mission and target community of the garden and everyday experiences in the place. Conventions intended to help foster community, paradoxically, functioned to erode the sense of community at the garden.

Overreliance on conventions can replace the vital process of collaboratively identifying the needs, strengths, and resources of the community the project (or product) is intended to serve or who stands to benefit most from it. As Becker suggests, conventions do not simply save producers from having to innovate, but they also stifle creativity.[52] This is especially true when a person or group *over*relies on them. Additionally, when producing something like a community garden, overreliance on cultural conventions thwarts the development of place-based embodied knowledge. Since conventions are necessarily modular, they tend to be nonplace-specific, and overreliance on them runs the risk of creating a disconnect between abstract social space and real, lived, and embodied places.

By adopting the linguistic and communicative conventions that aim to create "sustainable food systems" in Las Vegas, without collectively identifying what "sustainable" means, both organizationally and environmentally, in Las Vegas and HWLV, VRCG remains disconnected from the social, environmental, and cultural contexts in which it is situated. The same is true when using abstract concepts like the "local community" and the "neighborhood," that are not embodied in physical experiences.

Selective reliance on conventions can result in a doubly negative scenario in which the convention loses its time and resource-saving potential, and the cultural producer must get creative and innovative, but under constraints that result from the partial adoption of the convention. In this scenario, the cultural producer effectively makes the production process harder by establishing standards for which it only holds itself partially responsible. In deciding that VRCG would be a volunteer-driven, "user-created" place, and that outside "experts" would not be allowed to interfere with the vision for the garden, responsibility for mobilizing and/or constructing a community of "users" to create the place fell solely on Roz and her small group of volunteers. Breaking with the convention and paying day laborers to use their expertise to periodically maintain the garden space undermined the process of creating a sense of ownership of and responsibility for the garden by the "users."

Decisions about how much of a convention to adopt, or at what point to break with a given convention, involve issues of power. How power is distributed through an organization and who is granted legitimacy and authority to make decisions influence how conventions affect cultural producers' appropriation of broad cultural trends locally. I discuss organizational structure and the distribution of power at VRCG in chapter 5.

Urban places such as buildings, streets, neighborhoods, and gardens get reputations. These reputations exist largely in the imaginations of urban residents.[53] Initially, at least, the meanings associated with places emerge out of what people hear, read, and see about them. Cultural producers of urban places have control over what people see via their decisions regarding the physical construction of the site. They also control, again initially, what people hear about the place, since they are in a position to explain to the public who and what the place is for, in their own promotional materials and via the media. The impressions cultural producers try to give of the places they create can be understood as performances—as attempts to control the reputations of the place.

Overreliance and selective reliance on cultural conventions in the production process can create a disjuncture between the performance of place and the quotidian realities of the place. Audience members and "support personnel" alike tend toward "sign acceptance,"[54] and, at least initially, try to cooperate with the cultural producer to maintain the integrity of the performance of place. These actors engage in "rituals of avoidance," one of the types of "rituals of deference" described by Goffman, and avoid bringing up or challenging discrepancies or inconsistencies in the performance,[55] thereby trying to perpetuate the reputation constructed by the cultural producer.[56] Sometimes, however, the discrepancy between the performance and reality in everyday life leads to a "performance disruption" and a "spoiled" reputation of the place.[57]

NOTES

1. American Community Garden Association, "Find-A-Garden" (ACGA, 2019), https://www.communitygarden.org.

2. Lindsay K. Campbell, *City of Forests, City of Farms: Sustainability Planning for New York City's Nature* (Ithaca, NY: Cornell University Press, 2017); Alison Hope Alkon et al., "Introduction: Development, Displacement, and Dining," in *A Recipe for Gentrification: Food, Power, and Resistance in the City*, ed. Alison Hope Alkon, Yuki Kato, and Joshua Sbicca (New York: NYU Press, 2020).

3. Myrna Margulies Breitbart and Pamela Worden, "Creating a Sense of Purpose: Public Art and Boston's Orange Line," *Places: Forum of Design for the Public Realm* 9 (1994): 80–86.

4. Laura J. Lawson, *City Bountiful: A Century of Community Gardening in America* (Berkeley: University of California Press, 2005).

5. John R. Hall and Mary Jo Neitz, *Culture: Sociological Perspectives* (Prentice Hall, 1993).

6. McKelvey, "Community Gardening Toolkit."

7. Lyn H. Lofland, *The Public Realm: Exploring the City's Quintessential Social Territory* (Hawthorne, NY: Aldine de Gruyter, 1998); Borer, "The Location of Culture."

8. Melinda J. Milligan, "Interactional Past and Potential: The Social Construction of Place Attachment," *Symbolic Interaction* 21, no. 1 (1998): 1–33.

9. Howard S. Becker, "Art as Collective Action," *American Sociological Review* 39, no. 6 (1974): 767–76; Milligan, "Interactional Past and Potential." In this context, when I use the term "cultural producers" I am referring to Milligan's "set designers" or Becker's "artists," or those social actors responsible for developing the idea and plan for the cultural product since in general all social actors have the ability to create culture.

10. Becker, "Art as Collective Action," 771.

11. Becker, "Art as Collective Action."

12. Becker, "Art as Collective Action," 773.

13. Nina Eliasoph and Paul Lichterman, "Culture in Interaction," *American Journal of Sociology* 108, no. 4 (2003): 735–94.

14. Michèle Lamont and Virág Molnár, "The Study of Boundaries in the Social Sciences," *Annual Review of Sociology* 28 (2002): 167–95.

15. Becker, "Art as Collective Action," 722.

16. Or, to put it in Eizenberg's terms, I am examining how individuals draw from the institution of "community gardeners" to form a collective at a particular community garden. Efrat Eizenberg, *From the Ground Up: Community Gardens in New York City and the Politics of Spatial Transformation* (New York: Routledge, 2013).

17. Erving Goffman, *The Presentation of Self in Everyday Life* (New York: Anchor, 1959).

18. Gardens For All, *Gardens for All: A Guide to a Greener, Happier Community* (Charlotte, VT: Gardens for All, 1973); Mary Coe, *Growing with Community Gardening* (Taftsville, VT: Countryman Press, 1978); Boston Urban Gardeners, *A Handbook of Community Gardening* (New York: Charles Scribner's Sons, 1982); Lawson, *City Bountiful*. See Lawson, *City Bountiful*, 223–24, for a discussion of the role books played in spreading the community garden movement throughout the

1970s, 1980s, and 1990s. Specifically, she identifies how books provided laypersons with best practices, or, for the purposes of our discussion of cultural production, the *conventions*, of community gardening.

19. See for example the *Community Gardening Toolkit* created by Bill McKelvey (2009) of the University of Missouri Extension, which is a thorough guide adopted by cooperative extensions across the country.

20. Milligan, "Interactional Past and Potential."

21. Hilda Kurtz, "Differentiating Multiple Meanings of Garden and Community," *Urban Geography* 22, no. 7 (2001): 656–70.

22. McKelvey, "Community Gardening Toolkit."

23. Becker, "Art as Collective Action."

24. Isabelle Anguelovski, *Neighborhood as Refuge: Community Reconstruction, Place Remaking, and Environmental Justice in the City* (Cambridge, MA: MIT Press, 2014), http://ebookcentral.proquest.com/lib/utsa/detail.action?docID =3339756.

25. "Active" approaches to composting involve "turning," or mixing, the organic material and keeping freshly added matter separate from that which is further along the decomposition process. "Passive" composting is a more hands-off approach whereby the organic matter is stored in a pit or bin and largely left alone as it decomposes.

26. Francesca Polletta, "Culture and Movements," *The ANNALS of the American Academy of Political and Social Science* 619, no. 1 (2008): 78–96.

27. Polletta, "Culture and Movements."

28. Sarah Banet-Weiser, *Authentic(TM): The Politics of Ambivalence in a Brand Culture* (New York: NYU Press, 2012).

29. A Los Angeles–based group made up of individuals, small businesses, and organization that fought to lift municipal codes restricting the use of residential land for gardening and farming.

30. Alison Hope Alkon and Julian Agyeman, eds., *Cultivating Food Justice: Race, Class, and Sustainability* (Cambridge, MA: The MIT Press, 2011), 5.

31. McKelvey, "Community Gardening Toolkit."

32. Stephanie Tavares, "West Las Vegas Group Converts Vacant Lot of Land to Community Garden," *Las Vegas Sun*, March 22, 2010, http://lasvegassun.com/news /2010/mar/22/community-garden-sprouts/.

33. Lynnette Curtis, "Faith Makes Fruits, Vegetables Sprout on Five Acres," *Las Vegas Review-Journal*, July 19, 2010, http://www.reviewjournal.com/news/faith-makes-fruits-vegetables-sprout-five-acres.

34. Howard S. Becker, *Art Worlds* (Berkeley: University of California Press, 1984).

35. Goffman, *The Presentation of Self in Everyday Life*.

36. Goffman, *The Presentation of Self in Everyday Life*.

37. Goffman, *The Presentation of Self in Everyday Life*, 44.

38. Lawson, *City Bountiful*, 230.

39. Jeanette Abi-Nader et al., *Growing Communities Curriculum: Community Building and Organizational Development through Community Gardening* (College

Park, GA: American Community Gardening Association, 2001), https://www.communitygarden.org/product-page/growing-communities-how-to-build-community-through-community-gardening.

40. Laurie Harmon and Laurel Harrington, *Building a Community Garden in Your Park: Opportunities for Health, Community, and Recreation*, Grow Your Park Initiative (Ashburn, VA: National Recreation and Park Association, 2010), https://www.nrpa.org/uploadedFiles/nrpa.org/Grants_and_Partners/Environmental_Conservation/Community-Garden-Handbook.pdf.

41. Breitbart and Worden, "Creating a Sense of Purpose."

42. McKelvey, "Community Gardening Toolkit."

43. Diana Crane, *The Production of Culture: Media and the Urban Arts* (Newbury Park, CA: SAGE Publications Inc, 1992).

44. Goffman, *The Presentation of Self in Everyday Life*, 58.

45. Goffman, *The Presentation of Self in Everyday Life*, 58.

46. Goffman, *The Presentation of Self in Everyday Life*, 35.

47. Goffman, *The Presentation of Self in Everyday Life*.

48. Goffman, *The Presentation of Self in Everyday Life*, 56.

49. Goffman, *The Presentation of Self in Everyday Life*.

50. Becker, "Art as Collective Action."

51. Becker, "Art as Collective Action."

52. Becker, "Art as Collective Action."

53. Sarah Zelner, "The Perpetuation of Neighborhood Reputation: An Interactionist Approach," *Symbolic Interaction* 38, no. 4 (2015): 575–93.

54. Goffman, *The Presentation of Self in Everyday Life*.

55. Erving Goffman, "The Nature of Deference and Demeanor," *American Anthropologist* 58, no. 3 (1956): 473–502.

56. Zelner, "The Perpetuation of Neighborhood Reputation."

57. Goffman, *The Presentation of Self in Everyday Life*.

Chapter 5

"We Have Everything Else, but We Have No Foundation"

The Impact of Strategic Choices on Collective Identity Formation

Late one March morning, I arrived at the garden to an empty parking lot, but the garden stereo was blasting 1970s funk music. Jimmy always had the radio tuned to 88.1 FM, POWER 88: The Soul School Station, so I thought I might find him inside the fence. Jimmy was alone and watering seedlings at the green mesh hoop house. These were plants that would very soon be transplanted into the rows for the "You-Pick" farmer's market. Jimmy saw me, walked over, and yelled, "There he is!" He extended his fist for a bump. I asked about his weekend. He paused and said, "*Good*!" Then he said, "Today's *Wednesday*!" I said I knew that, but that I hadn't seen him since Friday. He laughed. I asked him if he thought the seedlings were ready to be planted. He let out a low laugh and said he was not going to plant a single seedling until the (holding up air quotation marks) "experts" told him to. I asked who the experts were, and he said, "I don't know. Not *me*. If *I* plant they'll blame everything on me if something goes wrong."

Jimmy was frustrated. He was going through a period of unemployment that allowed him to be at the garden nearly every day of the week. If it were not for his, often unpaid, dedication to keep the plants alive, there were numerous growing seasons when planting may not have happened on schedule, or at all, without him taking initiative to start seedlings and hand water them. There were also instances when personal or professional obligations arose that took him away from the garden and entire batches of seedlings died. Many volunteers I spoke with mentioned the critical role Jimmy played at the garden, acknowledging the fragility of the organization since it depended on his continuing unemployment. Peggy summed up this sentiment saying, "Thank God for Jimmy, but you know how long can that continue?" Jimmy's previous experience led him to wonder if he did not take responsibility for growing the seedlings and having them ready for planting who would? He was not so much frustrated that others were less concerned with this task—he enjoyed it, after all—but that if something went wrong once the plants were in the ground and people wanted to harvest food, he was the one to shoulder the blame if he planted them. In other words, the informality of the organization put him in a position where he was vital to the persistence of the garden, but he held little power.

The apparent order and control in the garden space of Vegas Roots belied a more unstable and capricious foundation. The dis-order at VRCG is largely interpersonal and organizational, leading to unpredictable experiences and interactions for some participants and visitors in the garden space. Ambiguities in the division of labor, distribution of power, and norms of decision-making complicated the process of developing a sense of unified vision, shared ownership, and collective efficacy among garden participants. There certainly were structural elements impacting the ability to mobilize physical, human, and social capital for a community garden in Las Vegas. In this chapter, however, I explore the agentic choices made by garden participants that constrain collective identity formation.

Many urban community gardens are both place-based and interest-based. That is, they attract interested parties from different parts of a city or region to a particular place where the food project is located. Although some gardens espouse a "for the neighborhood by the neighborhood" ethos, many do not restrict participation, outreach, or intended audience so narrowly. Much of the literature on urban community gardening conflates the concepts of "community" and "neighborhood," or assumes the communities of these urban places preexisted and could be mobilized or recruited.[1] Alongside these types of projects are ones better described as leisure-based "communities of acquaintances," which sociologist Gary Alan Fine suggests are localized types of "lifestyle enclaves."[2] Gardens built around interest-based, spatially unbound conceptions of community rely on "commuter" participants from outside the neighborhood and face unique obstacles in forging a sense of shared mission, vision, and purpose, or what social movement scholars call "collective identity."

Community gardens are both places and organizations. They are both meaningful locations (a garden) *and* the social actors that collectively cultivate and inhabit them (the community). Appropriating broader cultural trends of urban community gardens in a local setting involves not just the production of the place itself (see chapter 4) but also the construction of the social bonds that tie the gardeners together. Whereas social movements were once understood as the mobilization of existing commonalities (e.g., the working class), changes in late modernity have shifted attention to the need for constructing commonality in the mobilization process.[3] Commonality is no longer a given, but must be cultivated. Today, there is more emphasis on how collective action groups negotiate a collective identity, or "the processes through which a collective becomes a collective."[4] This involves an ongoing coconstruction of a shared bond that helps individuals make sense of what they are doing together. Like community, collective identities are ongoing accomplishments.

Community gardens have become a popular addition to many cities' sustainability initiatives in part because they have the potential to positively impact a constellation of social, economic, and environmental issues. Activists, politicians, and planners highlight their ability to contribute to many progressive urban ideals like "community food security; urban ecosystem health; active living; pedestrian-friendly neighborhoods and open-space networks; and the equity concerns of low-income and minority communities, immigrants, and seniors."[5] The diversity of causes that intersect at urban community gardens is typically framed as a reason to promote urban gardening. As important are the organizational complexities of juggling these different issues and the potentially divergent interests of participants attracted to urban community gardens by them. When a community garden draws participants from all over the city, as does VRCG, and from a diversity of separate "movements," causes, or other motivations for action, as discussed below, its organization must strategize about how to forge a shared sense of purpose. This process, even under ideal conditions, tends to be fraught.[6] *Who* is allowed to make decisions regarding these issues depends on some of the organizational strategies described below.

A collective action group's ability to achieve its goals is largely dependent upon the *strategic choices* they make regarding various organizational issues. As part of a larger cultural turn in social movement theory and research, James Jasper proposed a "strategic framework" to collective action, foregrounding the way groups make strategic choices to attain their shared goals. Strategic choices require an interpretive process through which a variety of courses of action are whittled down to one. When there is no clear "best" choice, which is often, groups are faced with a strategic dilemma.[7] This chapter explores the strategic choices Roz and various core volunteers made in the face of some of the dilemmas they encountered. I explore the impact of these choices on the process of collective identity formation among garden participants.

CHOICES IN COLLECTIVE ACTION

One area of collective action where we can locate and analyze agency is in the moments when groups encounter "strategic dilemmas" and face "two or more options, each with a long list of risks, costs, and potential benefits."[8] Since there is rarely one "right" answer in these scenarios, groups must make "strategic choices" about how to proceed based on perceived tradeoffs. "Choice" for Jasper simply means that one could have acted otherwise. Sometimes groups are hardly aware a choice has been made, but just because alternatives were not discussed, or perhaps even perceived, does not mean they were not available. Indeed, "individuals and groups must initiate or pursue one flow of action rather than another, respond in one way to events rather than another."[9]

Strategic action tends to take on a habitual character because most collective action occurs within "sequences."[10] A sequence is a path or a trajectory, often initiated by a strategic choice, which cascades and unfolds over time. Sequences can entail a degree of path dependency, with choices about which path to pursue rendering certain future choices more likely and others less. Emergent collective action groups develop patterns of action, and over time, those routines foreclose certain means and ends.[11]

In this chapter, I focus on sequences and turning points in the garden's history, paying attention to the ways they have shaped, and been shaped by, choices made to address two strategic dilemmas faced by participants at VRCG: (1) the *organization dilemma*, and (2) the *shifting goals dilemma.*[12] These dilemmas correspond with broad questions every nascent collective action group needs to answer: How should we treat each other? and What's the problem?[13] Answers to these questions form shared touchstones participants evoke to create a sense of coherence and continuity surrounding their collective action.

The strategic choices a collective action group makes regarding its organizational structure, the criteria for participation and membership, and the goals of collective action can have important impacts on the collective identity of the group. Collective identity has been described as "the shared definition of a group that derives from members' common interests and solidarities."[14] Simply put, collective identity can be understood as a sense of "we-ness."[15] Since interests, solidarities, grievances, and ideologies are not structurally determined, even among collective actors who share structurally grounded positions (e.g., members of a common social class, racial or ethnic group, gender, religion), collective identities are not latent shared definitions lying in wait to be "mobilized" or for "political opportunities" to open. They are social constructions.[16] Collective identity does not precede collective action.

Individuals engaging in collective action construct a bond because they need it to make sense of what they are doing together.[17] Not all people who share an ideology or identity with members of a collective action group join the group.[18] Not all who join are motivated to do so by the same goals, interests, and values. Accordingly, the process of collective identity construction involves "group identity work" through which group members create shared "symbolic resources" that serve as a bond.[19]

This chapter examines how VRCG participant addressed two strategic dilemmas and the impact their choices had on the process and practice of collective identity. Since strategic dilemmas can be interdependent and tend to overlap in the real world, this chapter also explores how strategic choices impact concurrent and subsequent strategic dilemmas. My observations suggest that choices members make to settle one dilemma influence the nature of other dilemmas and shape what sorts of choices members perceive as possible and appropriate.

The Organization Dilemma

The *organization dilemma* describes tradeoffs collective action groups encounter along two dimensions as they make choices about their organizational structure. They must decide how formal the structure of the organization will be, and they must decide if power will be held in the center or throughout the organization.[20] Some groups successfully achieve their goals through the creation of formal, bureaucratized structures,[21] while others find their most ambitious goals and effective tactics become less feasible as they go through the same process.[22] Additionally, collective action organizations can be thought of as existing along a continuum between centralized and decentralized.[23] There are costs and benefits associated with both formal and informal organization structures as well as centralized and decentralized distributions of power, so where a collective action group falls on these matters shapes the horizons of possibility, perceptions about tactical efficacy, and subsequent strategic choices.

Scholars of strategic collective action tend to favor certain configurations of the available options when confronting the organization dilemma. Some highlight the efficacy of centralized power in a hierarchical structure.[24] Others opt for decentralized power within a more loosely organized structure.[25] Still others, decentralized power within a "polycephalous" organizational structure comprised of various smaller hierarchically organized groupings.[26] Less discussed are scenarios in which centralized power is coupled with a lack of both organizational structure and clear division of labor. Or instances when power is decentralized in a hierarchically organized structure, for example, the reality behind the "holocratic" corporate model embraced by companies like Zappos.[27] The lack of attention to alternative responses to the organizational dilemma is due, in part, to a tendency to focus understandings of collective action on what has worked. This has led many scholars to study established groups and focus on organized activism, neglecting groups that are fledgling or less organized and the insights on collective action we might gain from them.[28]

Power Distribution: Centralized ⟷ Decentralized

For the first few months after the grand opening, Roz cultivated the garden space largely by herself. She had no core volunteers, yet, and very few garden volunteers in general. Roz remembered the first few months as a time of solitary struggle:

> Well, the first, you know after that first grand opening when we had like 80 some people there then we had like one person the next week. After we were in the RJ [*Las Vegas Review-Journal*] the first time, one by one people started coming in and saying, "Hi," and they heard about us and then you know I'd look around and there'd be three and four and you know so just little by little a little community of people started coming and helping.

This process was slow. Sarah, an early office volunteer and eventually a core volunteer, remembered the toll going it alone took on Roz: "I remember it just being hard on her. *Really* hard. Just the learning, and having a nonprofit. Just trying to figure [it all] out."

When I asked Roz about her vision and goals in the early days of the garden, when it was more of an idea than a physical reality, she was clear about her preference for action instead of talking and planning:

Roz: I didn't plan long term; it was just whatever happened from week to week. I didn't even try to make a plan; everything just evolved.
Tyler: What was the reason for taking that approach?
Roz: Because this is not my thing. You know, how do you plan something you don't even know? If I had done a community garden *before* somewhere then of course I would know . . . any other business, you know, I would know, but this was brand new to me so how to you make a plan? You know? You just plan to grow some food and see what happens.

This passage highlights two of Roz's core beliefs: she believes in the importance of learning by doing and that people tend to either make plans or take action; that planning is procrastination. These two beliefs intertwined in her approach to starting the garden. She explained, "I'm a firm believer, if you make the first step the universe will make the next fifty, and we've seen that over and over [at the garden]. But you have to make the step." Roz's reluctance to acknowledge that she could have acted otherwise, and established a formal structure or recruited volunteers with expertise in desert gardening and community organizing, *prior* to opening the garden does not mean that she did not make a strategic choice, and this choice initiated the earliest sequence of action at the garden. Her choice to "take the first step" alone established a model of centralized power resting solely with her. Describing working with Roz in the early days, Sarah summed it up saying, "It was her way or the highway."

Early strategic choices Roz made at the garden were past-oriented. Her initial approach to the garden was rooted in habit, reflecting past successful leaps of faith, like quitting her job as an elementary school teacher, attempting to start a homeless shelter, and starting her own mobile oil change company. Even though her homeless shelter project was eventually abandoned, it resulted in her being approached to start the garden. The garden was not her idea. She admits, "Yeah I would have never thought of a community garden, not ever." The genesis of the idea for the garden is important. Since it was a concept introduced *to* her, not *by* her, her vision for the future of the garden was limited. Roz's past experience with successful leaps of faith and her reluctance to make long-term, comprehensive plans before starting the project led her to practically evaluate present conditions and decide on an "if you build it they will come approach" approach of forging ahead in hopes of attracting others along the way.

As the "little community of people" grew, Roz and other volunteers attempted to establish a formal organizational model. After the garden was open a few months, more committed volunteers began to get involved in the project, and in May 2011 Roz formed an advisory board that was comprised of volunteers, myself included, who helped plan garden events, classes, and programs, worked toward creating a business plan and promotional materials, tried to assemble a board of directors, and strategized volunteer recruitment. The advisory board was a rotating cast of volunteers, most of whom had professional experience, and its creation marked a turning point, initiating a new sequence of action at the garden.

Sequences in collective action are complex, however, and their durability is always accompanied by a potential for dynamism. Collective actors have the ability to reinterpret their structural environments, reconsider appropriate means and ends, and disrupt the routinized actions of sequences.[29] There are moments in collective action when trajectories change, agency disrupts habitual actions, and the group faces a "turning point."[30] Turning points are shifts, or breaks, in meanings and interpretive frames shared by members of the group, which open the group up to the consideration of new possibilities.

The first meeting of the advisory board in April 2011 was focused on "infrastructure development," and the invitation contained a message from Roz stating, "The garden is growing by leaps and bounds and the media, groups, and volunteers are now all over us. I'm afraid that if I don't get some concrete things and people in place, I'm gonna' crack!" She acknowledged that as the garden grew, she would "crack" if she alone functioned as the foundation. She needed a more stable, "concrete" foundation of human and material capital to continue forging ahead.

The sequence of action initiated by the formation of the advisory board marked a shift toward hierarchy in the division of labor and organizational structure of the garden. Ideas from individuals besides Roz were considered and taken seriously, tasks of developing infrastructure, official documents, programs, and events were doled out to small groups led by volunteers. It was a dynamic time, organizationally, as Mary, a member of the advisory board, remembered:

> It shifted a lot. My sense at the beginning was that it was Roz and a cadre of very committed volunteers . . . Roz was the center of it. Roz was spending a lot of hours in the garden, but also had increasing demands to be out being the public face and building partnerships and she was doing a lot of both, relying on volunteers to help run the day to day to make sure things got planted, and things got weeded, you know, and pulled and picked, and managing volunteers.

Mary rightly described Roz as the nucleus of the group. Others were tasked with development projects, but Roz had the final say. Power remained centralized, but a more formal division of labor and approach to long-term planning began to take shape.

The more bureaucratized structure that developed via the advisory board provided volunteers with more freedom to utilize their expertise from outside the garden. Ensuring that volunteers would experience this potential as pleasurable demanded that their expertise be valued from start to finish, which was not the case for some volunteers, like Sarah. As she remembers:

> I could be tasked with things I was capable of doing, and I had a way I saw it should go; I had knowledge or expertise beyond her. But if it wasn't something she wanted, I couldn't move forward and so I'd be frustrated. I could help her—I could still be a part of it—but it wasn't really to the best ability that I could do or what I thought would be best for the project based off the expertise that I had. I could do something, but I couldn't take ownership because it wasn't something I believed in necessarily.

Sarah's use of the word "ownership" draws attention to the related issues of membership and belonging, two central elements underlying a group's collective identity. Once granted membership in a collective action group, participants expect, and are often obligated, to help shape the direction of the group.[31] She acknowledged she "could still be a part of it," but participation is distinct from collaboration. In collective action, decisions about "who belongs" are also claims about who the group or project *belongs to.*

Peggy, a plot renter and core volunteer, described a period of time while she was heading up a project as a member of the advisory board. Like Sarah, she described feeling like she was tasked with using her expertise to guide the project but then felt her autonomy was diminished:

> That was a season when I felt like I was owning it, and then at the last moment there were a lot of changes, and then I really wasn't sure what my role was any

> more, so then I backed off. In terms of leading that team. Like, am I the boss or not?

Sarah and Peggy describe feeling left out of the final decision-making process. Since the process of collective identity involves making sense of what collective actors are doing together, when those engaging in the hard work of building a grassroots organization neither experience a sense of ownership of the group nor are allowed to make decisions, the process of collective identity formation is thwarted. When belonging to a collective action group is detached from ownership of the group, some of the pleasures associated with participation dissipate. Grassroots participation is pleasurable largely because it gives social actors, the experience of feeling like their subjectivity and agency contribute to shaping the group, its goals, and the means through which it pursues them. Distinctions within a group regarding whose ideas are considered credible and listened to, and whose are not reflect hierarchies of power,[32] render participation less enjoyable for those getting ignored as they feel like they not valued for their ideas, only their willingness to pursue others' ideas.

Although the advisory board represented an attempt at a more hierarchical, bureaucratic organizational structure, the process was sometimes contentious, and the group's attempt at constructing an active and involved board of directors is an illustrative anecdote. This effort quickly disintegrated. During a meeting on February 16, 2012, one of the advisory board members working on forming a board of directors inquired about a perceived lack of interest in our recommendations from Roz and her avoidance of our questions about the existing board and their responsibilities. Roz responded she was afraid of losing her "baby" (the garden), explaining she knew people who started projects, recruited ideal board members, and were subsequently "voted out" by the board after disagreeing on the direction of the project. Specifically, Roz and the other core volunteer in our meeting were debating how much money to have prospective board members pay or raise. Roz kept referring to "selling the garden." Roz did not give in, the volunteer who tried to convince her to ask for less grew frustrated and stopped attending meetings, and the discussion of a board of directors was put to rest. By September 2012, the advisory board disbanded.

The preceding anecdote highlights a turning point at the garden, marking the end of one sequence of strategic action. While she wanted the division of labor that accompanies a more formal, bureaucratic model of organization, Roz resisted changes in the distribution of power that would accompany the establishment of a more official board of directors. In an interview a few months later, she explained her rationale:

> I won't get an official board who can give 'cause most board members are responsible for giving five or ten thousand dollars a year, but you know in

> exchange they have a lot of control. I need the garden to run organically. I need it to be able to just do whatever it does. Even I'm not trying to be so rigid in . . . I don't know what the hell controls nor do I care, um I just want it to be able to evolve. I can't have you telling me what it's going to evolve to. How the hell do *you* know? So consequently we've got to live off pennies.

Her opposition was largely due to her intense emotional response to a scenario in which there would be a potential for her to lose total control over the garden. Strategic action is often understood in terms of rational choices, which is typically positioned in opposition to emotions. Emotions, however, play an important role in agentic action and strategic choices.[33] In the following months, all the members of the advisory board drifted away, some out of frustration, others for personal reasons. Josh, who left for personal reasons, described a noticeable change in the organization when he returned:

> I had some fairly significant family issues happen so I missed a couple meetings . . . when I tried to reengage, I think at that point the committee was pretty much gone. . . . And then trying to reengage was kind of like, "Oh, this is very different than what I left even just a couple months before."

The disagreement over the board of directors altered the trajectory of the organization. It did not only turn off some volunteers, but it also led to a shift in the temporal orientation of strategic action at the garden. Prior to the advisory board, most strategic action was oriented toward present concerns and sometimes by past successes or failures.

The sequence set into motion by the establishment of the advisory board was initiated by a desire to replace the myopic temporal considerations informing previous decision-making with a concern for future sustainability of the garden. Once created, actions taken by the advisory board became increasingly characterized by intentional actions of a habitual nature. The group worked together, striving for, and sometimes achieving, consensus about goals and tactics, and it continued to do so because it had worked in the past. The dispute caused Roz to reconsider the groups' actions in light of the present situation. Since the issue at hand, the replacement of a place-holder board of directors with a real one, was one about which her opinion differed greatly from other advisory board members and about which she did not want to negotiate with the board. She feared if the organization continued to operate with a formal organization she could find herself in a scenario in which she would have to cede power and a board of directors could vote her out.

The formation of the advisory board represented a *potential* turning point, initiating a new sequence at the garden characterized by the establishment of a firm foundation and organizational infrastructure. Resistance to a formal board of directors and participants' perception of a lack of autonomy and shared decision-making authority thwarted such a shift. The dissipation of the advisory board was a turning point that started the garden down a new path, a sequence in which the external goals—what participants wanted to accomplish through the garden—remained unchanged while the internal goals—what the participants hoped to accomplish organizationally—greatly changed. Roz made the strategic choice to centralize decision-making authority solely back into her hands, taking on the responsibility of anchoring the garden by herself, a position she previously acknowledged was untenable.

The garden remained in the same sequence for about three years. Power continued to be centralized solely in Roz's hands, and as an organization, the garden remained loosely organized. This combination produced a sense of disorganization and confusion in day-to-day garden operations. There were ebbs and flows of garden volunteers, office volunteers, and core volunteers, resulting in inconsistent online communication and outreach and on-site planning and daily operations. Roz started the advisory board because she began to feel overwhelmed when the work of maintaining the garden was increasingly accompanied by demands related to running a successful nonprofit organization. The latter came to dominate more of Roz's time after the advisory board dissolved, but she continued to exercise total control over decisions made regarding daily operations in the garden. Frustration over the organizational form and centralization of power at the garden came up frequently in my interviews with volunteers who were active during this sequence.

One volunteer, Alex, described a lack of foundation at the garden and suggested the garden should change its structure to adapt to the desires of potential participants:

> I would really like us to be more in tune with the people who come here. We should have more hours, we should have more time to give. And plus all the people who do come here who eventually hope to be a part of this team, um, that that would be an option.

Specifically, she was referring to the fact that the garden is open mostly on weekday mornings, when many people are working. She wanted the garden to be more flexible in order to adapt to residents' schedules and other needs. She suggested there was a need for volunteers to have their voices heard and ideas considered. This sentiment was echoed by George, a long-time garden volunteer and recent core volunteer, when I asked him what he would change about the garden if he could:

> I would get a lot of input and listen to the volunteers that are here all the time. Because you may have one idea of what you want the garden to be, but you

> have all these consistent volunteers, and they see changes and know what needs to be done. I feel it's important to listen to that. Because they're here a lot, and they're the ones doing a lot of work so I feel that input is very important because sometimes I feel that they feel they may not be very appreciated.

George felt there was untapped expertise and creativity available at the garden. Whereas Roz tried to recruit volunteers who possessed skills that could help her pursue the projects and goals she developed, George believed those who volunteered regularly developed expertise about how the garden operates, what needed to be done, and how best to do it. Accordingly, he felt volunteers should not be recruited simply as laborers but also as collaborators.

George's comments, combined with Peggy and Sarah's above, point to a broader organizational trend that cut through the entire garden trajectory. Participants felt they were able to contribute their time and energy, but they could not shape the direction of the organization or garden project. Even when individuals were tasked with a project, they sensed that ultimately their decisions were subject to a veto by Roz. The dynamic is best described as micromanagement, and just like in a workplace, the result has tended to stifle creativity and innovation, make volunteers feel like their ideas and opinions are not trusted or taken seriously, and ultimately led to disengagement and contributed to a high turnover.

Organizational Structure: Formal, Bureaucratic ⟷ Informal, Grassroots

The grievances brought about by micromanagement are related to the centralized power at the garden, and they are accompanied by a perceived lack of a stable foundation, a concern that speaks to a loose organizational structure. Some volunteers described feeling like there is a lack of consistent norms, routines, policies, and rules, leading, at times, to unpleasant and/or confusing experiences for volunteers and visitors. Josh believed that the concentration of power at the garden rendered the informal organizational structure less effective and contributed to negative experiences for visitors and garden volunteers.

> I think organization is huge. I think when people go they don't necessarily feel like they know what they're doing or why they're doing it. And that was one of the first reasons I got involved, I was coordinating volunteers; that was my job. I was like, "Roz, I will help coordinate volunteers. I can't be here all the time during the week if somebody comes, but during the weekend I have no problem setting structures up to make sure volunteers know what they're doing, and why they're doing it, and why it's important that they're doing it, and help get them to the point where they're self-sufficient and they don't need me to tell them

what to do, or need *anybody* to tell them what to do. They can just come and do it. So I think the disorganization is the most difficult thing.

For Josh, if the garden was to rely on volunteer labor, there was a need to establish some order in the day-to-day activities in order for volunteers to understand how their work contributed to the larger goals of the organization. He was expressing a desire to create a "meaningful, shared ground of interaction," or what.[34] Group styles help filter the collective representations social actors use to make interactions meaningful. This highlights the importance of collective identity in collective action, since it is not just the core volunteers who need to construct a bond in order to make sense of what they are doing together but also the desire of temporary volunteers to tap into that bond during their more fleeting experiences at the garden.

As Josh elaborated on this point, he also suggested that the potential versatility and adaptability of an informal organizational form was thwarted by the concentration of power and decision-making authority solely with Roz. As a person with volunteer coordination experience, he wished core volunteers had been able to have more influence over decisions about when to accept volunteer groups and how to prepare for them. He explained what he thought was missing:

Some sort of structure so it's not everyone going to Roz, so when you get to the garden and you want to volunteer there is a staff person there. And understanding if we cant provide staff five days a week then we need to make sure we're not opening the garden to volunteer groups on days when we cannot provide someone to be there. There's a point when you have to tell groups, no, we cannot accommodate you on that day. I remember having a conversation, we had agreed Tuesdays and Saturdays we could have volunteer groups, but then Roz started accepting volunteer groups on days that weren't Tuesdays and Saturdays. And it's like, welp, I mean those are the days we can realistically manage groups. Every other day we cannot do it because there's no one there. And I don't expect Roz to be there to manage those groups, it really is up to other people, but if those people cannot be there we shouldn't be having volunteers on those days.

The centralization of power and decision-making authority led to interactions between the garden organization and the public that failed to capitalize on the dedication and expertise of the loosely organized skilled volunteers. If there was work to be done at the garden, Roz would not turn volunteer groups away. She valued the amount of work that could get done, whereas Josh believed the meaningfulness of the labor to the volunteer was more important. When Roz began accepting volunteer groups whenever they requested to visit, she often had to coordinate and manage them herself since other core volunteers were not always available. I personally received texts regularly throughout 2012 and 2013 inquiring if I was available to be at the garden to manage last-minute volunteers. I often was not, and she would end up having to do it herself while also juggling meetings, public presentations, correspondence with media, volunteers, and donors, and other administrative duties associated with running a garden and nonprofit organization.

Accordingly, social media, e-mail, and phone correspondence was inconsistent. Volunteers were often put to work without explanation of the impact of their actions. Instead of simply being put to work, Josh believed garden volunteers would have had a better experience if they also learned about horticultural processes and what would be done with the food once harvested. Roz was often too busy to provide such a thorough volunteer experience, and at times she relied on new garden volunteers or people doing community service. Alex, a core volunteer who started out at the garden completing court ordered community service recalled one of her early experiences at the garden:

> On my second day I was already supposed to be able to explain the garden, I was still doing community service here, so I was just doing work. When people came in, I was supposed tell them what the plots were, where the rows are, about the chickens. Though I have the basic thought of it in total because I've done things like this in my life, it's not the same. It was sort of like, *what*?

Alex did not feel informed enough herself to provide quality explanations of the garden's origins or the practices and programs like the adoptable plots, pick-it-yourself market, the hens and their eggs, among other questions visitors may have. She did not feel adequately prepared to take on such responsibilities.

In another instance, she was shouldered with responsibilities at the garden but not given authority to make decisions. As she described, this situation created an unpleasant experience for both her, as a core volunteer, and a plot renter:

> I had somebody complain about her irrigation lines. I was like OK, I'll totally change that for you. One, her valve wasn't on. Nobody checked to see if her valve was on. So she had seeds in the ground that were getting no water except for the two days that she came out of the week. Nobody turned over her soil or added compost. Whenever somebody buys a plot that's what I do so it's ready to go. She asked me to change [her irrigation lines], she had the old tapes. I was like, yeah I totally will, you should have the new ones. Then when I went to do it, "Oh no, we could save money by not changing it." I didn't change 'em

> because I was told not to, even though I had already given my word. That's frustrating. You have somebody telling you we're going to do this for you. Jimmy and I. And then you have somebody telling 'em no, they don't need [new lines]. Office. A and B. You can't have A and B.

Alex was frustrated that her attempt to rectify an unpleasant experience for a plot renter was vetoed after she had promised new drip irrigation lines to the woman. She felt the cost of the new lines was less important that ensuring the plot renter had an updated plot. She also used the experience to draw attention to a rift she believed existed between the people who do most of the labor and interacting with volunteers and visitors at the garden, what she called "A," and Roz, or the "office, what she called "B."

Alex elaborated on the divide between A and B, explaining that she wished there were more dialogue between the two:

> We have no foundation. We have everything else, but we have no foundation. [Ideally] we'd have people here, office would be open, everybody knows the combination to the [tool] shed if they come here often enough. . . . We're kind of separated. I do what [Roz] says and I and then I do what Jimmy says. And when they come together they don't want to do the same thing. I think if we had a group that could come to consensus it would run so much better. 'Cause we'd be on page. And when people come in, when people start to notice the tear it gets weird. We should be one, not A and B.

Her comments echoed an issue I heard repeatedly among core volunteers: their sense that the people who labored the most in the garden had developed knowledge about the place that should be included in decision-making processes.

George outlined what he thought Roz needed to do in order to mend the rift:

> Listen to the volunteers that are here all the time. Because, OK, you may have one way, one idea of what you want the garden to be. But you have all these consistent volunteers working and they see changes and know what needs to be done. I feel it's important to listen to that. Because they're here a lot more times than usual. And they're the ones putting their hands into it and doing a lot of work so I feel that input is very important. Because they have ideas and they do a lot of the work and sometimes I feel that they feel they may not be very appreciated. And when you don't do that and when they come out here and you kind of tell 'em "do this, do that" I feel, I don't know I feel like they're not appreciated. And then you start to notice they show their face less and less.

Importantly, George was not saying an informal structure was the problem. Instead, he was suggesting the distribution of decision-making authority should be altered to better utilize volunteer expertise and knowledge and to help volunteers feel a bond with the project and others involved. He elaborated on how he thinks this process could happen:

> I'd start up some committees, also. Kind of give some people some responsibilities and instead of having everybody try to work on the same idea or push one direction, you get em all together and you put em in charge of certain things. (George)

He was describing an organizational structure that was segmented, with power divided among volunteers, and decision-making authority shared with those "putting their hands into it and doing a lot of work."

The shifting responses to the organization dilemma trace different sequences on the garden's trajectory. Roz started without a plan, and in fact, spent the first few months almost entirely by herself in the garden. A little over a year after the garden opened, Roz formed the advisory board, made committees within it, and members tried to establish a more formal organizational structure. At an advisory board meeting in February 2012, Roz commented, "It's time to elevate the garden to the next level," and went on to explain the infrastructure we were working on would make it possible for the garden to reaching more ambitious goals. A few months later, the advisory board disintegrated, and the garden reverted back to a loosely organized group of volunteers, with Roz making all of the decisions on her own.

The garden did not reap the benefits of the formalized, bureaucratic, centralized organizational form or the segmented, decentralized form.[35] Instead, the garden's organizational form morphed somewhere in between and at times suffers from unclear chains of command and the appearance of disorganization to newcomers, which can be a barrier to volunteer recruitment and retention.

The Shifting Goals Dilemma

The *shifting goals dilemma* centers on the question of when to refocus group resources to pursue a different goal. Sometimes a shift occurs when a group decides the means they have employed to reach a certain goal have proven ineffective, and participants decide to strive toward a more modest goal they are more likely to reach with established means. Other times, when a goal proves difficult to achieve, groups scrap both the means and the ends and shift to pursue both new ends and means.[36] In still other instances, changes within the group or external to the group expand the horizons of possibility, making new goals seem achievable.

Serving the Neighborhood?

Initially, the goals for the garden were modest. When I asked Roz to describe her initial plan for the garden she explained, "I was just gonna' grow some food for the neighborhood." On the first garden website, she posted an entry stating, "The week before the groundbreaking, we passed out 400 flyers in the community surrounding the garden. A huge thank you to Donna, her son, Devon, Charles and Martin.[37] We covered a lot of ground in just a little time!" Remembering similar early efforts, Jimmy shared, "We used to have kids that come over that want to volunteer we'd give em' flyers and go tell em' to put em' on people's doors and cars all in the neighborhood." I spent a few afternoons with other volunteers handing out flyers at Pearson Community Center in HWLV. This goal of serving the neighborhood was picked up by local reporters covering the garden. In March 2010, an article in the *Las Vegas Sun* highlighted Roz's goals of helping low-income West Las Vegans learn how to grow food and incorporate fresh produce into their regular diets because of high rates of diabetes and obesity in the neighborhood. Later that summer, an article in the *Las Vegas Review-Journal*, like the previous article in the *Sun*, focused on the lack of access to food in the Historic Westside neighborhood as a reason Roz started the garden.

Some of the volunteers who got involved early in the project were motivated to do so by the mission to reach neighborhood residents. Josh recalled his initial impression of the organization:

> I thought we were really working on building something that would help feed people specifically in that neighborhood, teach people all over the community how to grow their own food, and create a sustainable organization that was selling some food to the greater Las Vegas community but also providing it specifically to the Westside community. (Josh)

But over time, discussions at meetings focused less and less on the neighborhood specifically. Efforts to draw in the neighborhood had been largely ineffective. If volunteers were to continue to dedicate energy to neighborhood recruitment, it would take a full-court press, but the advisory board had just formed, and issues of internal organization and fundraising began to eclipse neighborhood outreach.

George was disappointed that more of an effort had not been made to reach the neighborhood residents:

> I feel there's not as much promotion as there should be. There's not much community outreach. I know they told me maybe once or twice they had somebody

> knock on some doors and give people flyers. I was told they were mostly Spanish speakers, and of course everything is in English so they don't know what's going on. That's just once or twice in five years. That's not very effective. That's not very much outreach whatsoever.

He was drawing attention to a strategic choice. Roz and the advisory board had *refocused* nearly all resources and attention from reaching the neighborhood to focusing on building organizational infrastructure and creating garden classes and programs to raise money for the daily garden operations. In shifting the goals, the target demographic expanded and was less focused and defined.

The shift in goals was reflected in Roz's decision to change the name of the garden in early 2012 from Tonopah Community Garden, which is the name of the street where the garden is located, to Vegas Roots Community Garden. At the time, the advisory board was making headway in creating organizational stability and Roz's goals started to grow in scope. Strategic choices to move toward more formal organizational structures expanded Roz's perceptions regarding what was possible to accomplish. After a short break for the holidays, the advisory board met in early 2012, and at one point Roz said the organization was, "Not about the garden anymore. It's about *so* much more." She added that in the next five years, she wanted to have helped create fifty more gardens throughout the city and to "put fast food chains out of business." Part of the motivation for the name change was a broad plan to start a movement in Las Vegas centered around the notion of convincing Las Vegas residents to settle down, put down "roots" in the city, and start building up the public sector. Symbolically, the name change distanced the garden from the surrounding neighborhood and established a connection to the city in general.

Educating the City

Changing the goals of group or organization can cause a sense of discontinuity and discomfort for existing members. Josh characterized his experience on the advisory board as disorienting, due in part to changes in the goals of the group:

> It seemed like often we would switch our focus, and, to be frank, often times I thought the committee was on one page and Roz was on a completely different page. And the committee as a group all agreed on moving in one direction, but then because it was really not, well it was our organization in the broad sense of the word it was still Roz was at the head so it was kind of like what she said goes. So often times we would have to switch our focus or switch our mission depending on what her desires were or what she felt we had the ability to do.

Strategic choices regarding the centralization of power at the garden impacted subsequent choices about which goals to pursue and when to redirect resources toward new ones.

Shifting goals does not just disrupt the flow of collective action; it can result in disengagement if the shift is sufficiently distant from previous goals that motivated participants to action. Josh described one specific shift that made him question his allegiance:

> Roz was shifting from the educational and serving the Westside community focus to how to eat healthy and clean and focus on individual health. And while that is something I feel is really important that's not why I got involved in the garden, that's not my passion, that's not something I want to dedicate a lot of my time to because I feel like that's a personal decision, I don't really want to force that on other people. I want to educate people about the benefits of growing their own food, but I don't want to say, "this is how you must eat." And I kind of felt that that's kind of where we were going and that's not a direction that *I* could subscribe to.

During this period, the garden began hosting cooking events that promoted raw and vegan diets. Roz led groups through juice cleanses and detoxification diets. The garden's social media profiles reflected these shifts, too, dedicating more and more space to these lifestyle diets.

The goals did not simply shift over time; they also accumulated. After the advisory board disbanded and Roz took over planning duties again, her goals did not retract. She continued to take on new projects, volunteer groups, and community partnerships. She helped start a bi-weekly "marketplace" at the garden where people sold, for example, handmade jewelry, essential oils and soaps, clothing, and succulent plants. She helped start a monthly micro-grant event called Vegas SOUP, where local entrepreneurs pitched ideas for community projects and audience members paid $10 for a bowl of vegan soup and bread and voted on their favorite idea, and all the money raised went to the winners. There were also camp outs at the garden and numerous large volunteer events.

Randy's experience volunteering at the garden led him to conclude that the garden had expanded its lists of goals too broadly, and the organization was only able to superficially address each. In an interview, he emphasized a principle he saw as critical for successfully accomplishing goals:

> Working within the means that you have. *Scaling back* if necessary. That's my feeling. See a small portion of something that really works successfully, and then allow for a slow expansion after that.

Randy felt the garden had bitten off more than the organization could chew. At the tail end of the advisory board's existence, Mary, a board member, tried to help influence it in a manner that echoed Randy's ideas about acknowledging limits and expanding only once prior efforts had been accomplished:

> I advocated for narrowing the short-term priorities, like saying what are we doing and what are we not doing. It just felt like there was this tendency to say "we can do this *and* this" and sometimes being strategic means you're saying we're doing this and *not* this right now.

This strategy of paring down goals in order to focus meaningfully on a few attainable ones was not chosen, and throughout 2013 and 2014 the pace of activity at the garden slowed, plot rentals dwindled, newsletters were sparse, and Roz's reliance upon day laborers increased.

From Survival Mode to Rebirth

When things slowed down at Vegas Roots, Roz again made a strategic choice to redirect resources away from program and activity planning toward making sure the garden's bills could get paid. This has been a persistent concern since the garden opened, but it took on renewed importance during this period. E-mail blasts to past donors, volunteers, and newsletter recipients frankly stated the garden needed money in order to stay open and requested donations. One day in the spring of 2013, I ran into Roz at the garden, and she told to me there was only enough money to cover overhead financial costs for a few more months. She was upset but admitted that she was exhausted and that maybe this was God telling her it was time to call it quits.

Things stayed this way until late summer 2015. I was still visiting the garden, but my participation had dwindled significantly. Roz started holding "Vegas Roots Think Tank" sessions where volunteers could come and help brainstorm ideas for fund raising, programs, and events. It was similar to the types of meetings that started the advisory board years before. Soon there were monthly "Lil' Roots" classes for Las Vegas youth (and their parents) that taught gardening skills, introduced kids to healthy, fresh food, and facilitated arts, crafts, and exposure to nature in the city. The social media presence picked up again, too. In the fall 2015, Roz was awarded with two grants, one from the Federal Home Loan Bank of San Francisco (FHLBSF) and another from the U.S. Department of Agriculture (USDA), to finance a "Veggie Buck Truck." The VBT was intended to be a mobile farmers market and provide job training for young adults as well as serve an educational function.

Once again, just when it looked like the garden might not be able to be financially viable, Roz and volunteers initiated a new sequence that changed the course of the garden's trajectory, and their actions began to reflect a new, future-oriented temporal orientation.

POWER, STRATEGIC CHOICES, AND COLLECTIVE IDENTITY

Strategic dilemmas overlap, and the strategic choices that collective action groups make are interdependent. A strategic choice can initiate a sequence that takes on a path dependency that shapes subsequent choices. Choices about whose ideas are listened to carefully, and whose are ignored, establish distinctions between members that quickly become unmentionable.[38] In this chapter, I have illustrated specifically how choices made to address the organizational dilemma shape choices made with regard to the shifting goals dilemma. My field research suggests choices about the organizational structure shaped other choices, too, like those about whether or not, and to what extent, to adopt cultural conventions, outlined in chapter four and some that will be examined in chapter six that address the "extension dilemma."[39]

One of the first strategic choices made at the garden was for Roz, the director, to set out to create a community garden on her own, centralizing power. From that point on, power and the related authority to make decisions and veto others' has rested largely if not solely in her hands. The effect of centralized power on subsequent strategic choices has changed over time. The brief period during which Roz and the advisory board members worked to create organizational infrastructure and a more formalized division of labor resulted in some increased productivity administratively and in terms of fundraising and program development. Ultimately, disagreements about the organizational structure that would impact how power was distributed at the garden caused the group to dissolve and the garden to revert back to a more informal, disorganized organizational structure dependent upon a balance of individuals and small groups of volunteers, day laborers, and large corporate volunteer days.

Volunteers experienced the combination of informal organizational structures and centralized power that has been the norm for much of the garden's history as confusing, frustrating, and at times unfair. It is critical to understand that these feelings did not arise immediately for most volunteers or at all for some. Some were motivated to volunteer specifically to help with whatever needed to be done. They were not interested in guiding the direction of the project. It was when volunteers offered, or were asked, to take on larger projects that the combination of informal organization and centralized power might get in the way of them feeling like they could "take ownership" of their work and the garden.

My observations of these scenarios while in the field, and the sentiments expressed to me in interviews, were supported by speech patterns of volunteers in informal conversations. Many volunteers spoke about "them" and "they" when describing the garden rather than "we" and "us" language, as one might expect. This discursive pattern is a form of "speech norm"[40] among plot owners and garden volunteers that implicitly evokes a shared assumption about their connection to the place and a failure to instill a sense of ownership in volunteers and plot owners. This pattern is a discursive confirmation of some of the observations above, like Sarah's claim that she could not "own" the work she was doing for the garden and Alex's description of the garden as divided between A and B. People experienced their actions as disconnected from the larger project.

In addition to being a process, collective identity is also a practice, meaning it is in need of being reevaluated, reinforced, reformulated, or rejected. Even within collective action groups with seemingly clear and concrete collective identities, "what appears as a given reality, something more or less permanent, is always the result, at least to a certain extent, of an active process which is not immediately visible."[41] Accordingly, elements of collective action that presuppose the existence of a collective identity, like organizational structures, criteria for membership, and group goals, only indicate its potential existence.[42] The presence and contours of a group's collective identity must be empirically observed and understood as unfolding and dynamic. It is hard work to create a sense of collective identity, but as this chapter illustrates, they also tend to be fragile and difficult to maintain over time.

NOTES

1. Donna Armstrong, "A Survey of Community Gardens in Upstate New York: Implications for Health Promotion and Community Development," *Health & Place* 6, no. 4 (2000): 319–27; Troy D. Glover, "Social Capital in the Lived Experiences of Community Gardeners," *Leisure Sciences* 26, no. 2 (2004): 143–62; Karen Schmelzkopf, "Urban Community Gardens as Contested Space," *Geographical Review* 85, no. 3 (1995): 364–81; Shinew et al., "Leisure Spaces as Potential Sites for Interracial Interaction."

2. Robert N. Bellah et al., *Habits of the Heart: Individualism and Commitment in American Life* (New York: Perennial, 1985); Gary Alan Fine, *Morel Tales: The Culture of Mushrooming* (Urbana: University of Illinois Press, 2003).

3. Steven M. Buechler, "New Social Movement Theories," *Sociological Quarterly* 36, no. 3 (1995): 441–64; Blee, *Democracy in the Making*.

4. Melucci, *Challenging Codes*.

5. Jeffrey Hou, Julie M. Johnson, and Laura J. Lawson, *Greening Cities, Growing Communities* (Washington, DC: University of Washington Press, 2009).

6. Eizenberg, *From the Ground Up.*

7. James M. Jasper, *Getting Your Way: Strategic Dilemmas in the Real World* (Chicago: University of Chicago Press, 2006).

8. Jasper, *Getting Your Way*, 1.

9. Jasper, "A Strategic Approach to Collective Action: Looking for Agency in Social-Movement Choices," *Mobilization: An International Quarterly* 9, no. 1 (2004): 1–16, 2.

10. Blee, *Democracy in the Making.*

11. Blee, *Democracy in the Making.*

12. Jasper, "A Strategic Approach to Collective Action"; Jasper, *Getting Your Way.*

13. Blee, *Democracy in the Making.*

14. Leila Rupp and Verta Taylor, *Drag Queens at the 801 Cabaret* (Chicago: University of Chicago Press, 2003); Verta Taylor, "Mobilizing for Change in a Social Movement Society," *Contemporary Sociology* 29, no. 1 (2000): 219–30.

15. Alberto Melucci, *Nomads of the Present: Social Movements and Individual Needs in Contemporary Society* (Philadelphia, PA: Temple University Press, 1989); Verta Taylor and Nancy E. Whittier, "Collective Identity in Social Movement Communities: Lesbian Feminist Mobilization," in *Frontiers in Social Movement Theory*, ed. Aldon D. Morris and Carol M. Mueller (New Haven, CT: Yale University Press, 1992), 104–29.

16. Buechler, "New Social Movement Theories."; Steven M. Buechler, *Social Movements in Advanced Capitalism: The Political Economy and Cultural Construction of Social Activism* (New York: Oxford University Press, 1999); Melucci, *Challenging Codes*; Francesca Polletta and James M. Jasper, "Collective Identity and Social Movements," *Annual Review of Sociology* 27, no. 1 (2001): 283–305.

17. Melucci, *Challenging Codes.*

18. Sheldon Stryker, "Identity Competition: Key to Differential Social Movement Participation?" in *Self, Identity, and Social Movements*, ed. Sheldon Stryker, Timothy J. Owens, and Robert W. White (Minneapolis: University of Minnesota Press, 2000), 21–40.

19. Michael L. Schwalbe and Douglas Mason-Schrock, "Identity Work as Group Process," *Advances in Group Processes* 13 (1996): 113–47; David A. Snow and Doug McAdam, "Identity Work Processes in the Context of Social Movements: Clarifying the Identity/Movement Nexus," in *Self, Identity, and Social Movements*, ed. Sheldon Stryker, Timothy J. Owens, and Robert W. White (Minneapolis: University of Minnesota Press, 2000), 41–67.

20. Jasper, "A Strategic Approach to Collective Action."

21. William A. Gamson, *The Strategy of Social Protest* (Homewood, IL: Dorsey Press, 1975).

22. Frances F. Piven and Richard Cloward, *Poor People's Movements: Why They Succeed, How They Fail* (New York: Pantheon, 1977).

23. J. Craig Jenkins, "Resource Mobilization Theory and the Study of Social Movements," *Annual Review of Sociology* 9, no. 1 (1983): 527–53.

24. Jenkins, "Resource Mobilization Theory and the Study of Social Movements."

25. Luther P. Gerlach and Virginia H. Hine, *People, Power, Change: Movements of Social Transformation* (New York: Bobbs-Merrill Company, 1970).

26. Luther P. Gerlach, "Movements of Revolutionary Change Some Structural Characteristics," *American Behavioral Scientist* 14, no. 6 (1971): 812–36.

27. All organizations, social movements, grassroots groups, and corporations alike, face some version of the organizational dilemma. The "holocratic" approach is a corporate model in which there is a highly articulated hierarchy, but various spheres within the organization operate democratically and flexibly, for better and for worse, and those in charge are able to both take credit for good ideas and shirk responsibility typically associated with power and leadership.

28. Blee, *Democracy in the Making.*

29. Blee, *Democracy in the Making.*

30. Blee, *Democracy in the Making.*

31. Blee, *Democracy in the Making.*

32. Blee, *Democracy in the Making.*

33. Jasper, *Getting Your Way.*

34. Eliasoph and Lichterman, "Culture in Interaction."

35. Gerlach and Hine, *People, Power, Change.*

36. Jasper, *Getting Your Way.*

37. These are pseudonyms.

38. Blee, *Democracy in the Making.*

39. Jasper, *Getting Your Way.*

40. Eliasoph and Lichterman, "Culture in Interaction."

41. Melucci, *Challenging Codes,* 72.

42. Melucci, *Challenging Codes.*

Chapter 6

It's for Everyone, It's *for* No One

Explicit Inclusivity, Implicit Exclusivity, and the Boundaries of Community

Vegas Roots is promoted as the first *public* community garden in Las Vegas. On its face this claim seems clear enough. The garden is not for students at a specific school, members of a certain religious congregation, or residents of a single retirement community. It is open to the public. Upon further inspection, however, "public" becomes a more tenuous concept with important implications for how community is realized.

The ideal of public space can be roughly summarized as space that is open and accessible to all.[1] Scholars and activists have taken issue with this ideal in recent decades, pointing out that in practice public space is rarely open and accessible to *all* people.[2] Namely, there is always a degree of exclusion and control at play in public space.[3] Purely public space, locations that "are accessible to all groups and provide for freedom of action but also for temporary claim and ownership,"[4] is a rare phenomenon. The public pool, shopping mall, city park, urban sidewalk, and local coffee shop are just a few examples of places one might consider public, but are not truly open and accessible to all and certainly do not provide for "freedom of action." Despite often being controlled by private interests, locations like these are often experienced as public space by those who use them.

There exists an important yet often overlooked distinction between "public" and "community" places. Public often refers to the *ideal* of open inclusivity, while community refers to a balance of inclusivity and, importantly, exclusivity.[5] Architect and planner Alex Krieger explains,

> While we use the word "community" interchangeably with the word "public," community involves selection; a distinguishing of those who belong from outsiders. The public, on the other hand, is—or should—encompass everyone.

> This is not a subtle distinction, yet difficult for Americans whose ideals waver between demands for equality of access and territoriality.[6]

This distinction is crucial in discussions of food access, urban agriculture, and food justice. Although alternative food projects like farmers markets and community gardens have garnered support in a diversity of communities throughout the United States, food justice proponents understand that certain neighborhoods, certain racial and ethnic groups, and certain social classes are disproportionately in need of alternative food projects. These areas and populations also wield less power, influence, resources, and time than their more privileged peers, and accordingly, alternative food projects intended to serve these communities must be designed and operate in ways that take these factors into consideration if they are going to reach and resonate with their intended audience. This is especially true in cities or regions where alternative food projects are few and far between, like Las Vegas. A critical question for food justice-oriented community gardens is *which* community they aim to serve. A *public* community garden on the surface can sound as though it is inviting more people to the table, but although the ideal of a public place centers on notions of open inclusivity, in reality, public places often entail a degree of exclusion, even if it is unintentional.

This chapter contributes to a growing literature on race and alternative food. Located in Historic West Las Vegas, a predominantly low-income black neighborhood, Vegas Roots began with a mission to empower neighborhood residents and provide access to cheap produce, and ultimately became an expensive urban garden used by predominantly white suburbanites and corporations interested in boosting their public profile by supporting a community garden in a low-income predominantly black neighborhood. The findings illustrate how well-meaning food justice programs can lose their mission and fall into the trap of reinforcing the hierarchies they initially intend to resist.

COMMUNITY AND PLACE

Community gardens are places where people negotiate the meaning of community.[7] In this chapter, I explore how Roz and core volunteers define and talk about community and show how they engage in *boundary work*[8] to identify who is part of the "community" and who is not. I use one of the strategic dilemmas described by James Jasper, the *extension dilemma*, which involves deciding how inclusive to be in terms of criteria for individual membership as well as collaboration with other groups.[9] I am interested in how strategic choices regarding the symbolic and spatial boundaries of community at VRCG influence the manner in which the garden organization conceptualizes and interacts with the "community." Since the garden is a physical site where people interact and on behalf of which they interact, and since this place is located in a broader urban context, it is important to grasp how participants conceptualize the place itself and its relation to its surroundings when establishing and maintaining the boundaries of community. As a self-proclaimed community institution, whether or not, or to what extent VRCG acknowledges, appropriates, or addresses the history and reputation of the HWLV neighborhood affects how neighborhood residents perceive the garden and how participants from outside the neighborhood interpret the neighborhood.[10]

Instead of defining community and then measuring the extent to which VRCG meets that definition, I am interested in how participants define it and experience it. I examine how their conceptions of community address the issues of territoriality, social ties and networks, meaningful interactions (or community as practice), commonly associated with community. When a group sets out to create not just a garden, but also a *community* garden, from what operational definition of community do they proceed? What sorts of actions are produced by the definition? How does the definition influence participants' assessments of the state of the project?

PUBLIC AND/OR COMMUNITY

Shepard and Smithsimon offer a definition of public space that incorporates locations that approximate the public ideal as well as those places that are more controlled and exclusionary but some experience as public:

> Public spaces are places in which a range of people can interact with other people they don't necessarily know and in which they can engage in a range of public and private activities—though both the users and uses are inevitably limited.[11]

Although this definition is useful for grasping the role of exclusion in public, it reinforces the confusion between "space" and "place" within urban sociology. "Places" are physical and material sites with specific geographic locations that are invested with meaning and value.[12] Building from their definition of public space, Shepard and Smithsimon develop a typology of public "spaces,"[13] most of which would be more appropriately described as "places." Despite our definitional disagreement, their descriptions of the different levels of exclusion and sources of control that characterize public places can be useful for exploring how a community garden negotiates the tricky terrain between the ideal of a public place and the demands of a community place.

One type of public place[14] in Shepard and Smithsimon's typology is the "community place." These are places where exclusion is selective and the users of the place, not private owners or the government, exercise control.[15] They describe community gardens as exemplars of this type of place since those who are part of the garden organization are responsible for deciding who gets to participate. As the quote from Alex Krieger above indicates, although "community" is often considered synonymous with the ideal of the inclusive "public," there is always a degree of exclusion. Even when community places strive toward total inclusivity, there is a tendency to exercise some degree of exclusion, either intentionally or not.

Exclusion and control are not always explicit and direct; they can also take on more implicit forms that make some potential users feel uncomfortable or unwelcome. For example, consider how a café that does not post item prices, a park with "bumproof" benches,[16] a shopping plaza with exclusively high-end stores, or any location "for paying customers only" may deter certain individuals or groups from occupying the place. Additionally, consider how a cul-de-sac or small park can serve as a great place for neighborhood residents to congregate and play, but might also be uncomfortable, or perhaps even inhospitable or unsafe for an outsider. Public places are always subject to a degree of exclusion, both explicit and implicit, intentional and unintentional.

There were two dominant trends in boundary work at VRCG: *explicit inclusivity* and *implicit exclusivity*. The explicit inclusivity at the garden reflects what Julie Guthman identified as "universalizing impulses" within the alternative food movement.[17] Explicit attempts to avoid excluding any groups within the Las Vegas valley focused less on reaching low-income racial and ethnic minority residents in the neighborhood surrounding the garden and more on reaching residents throughout the valley to communicate, "Yes, this is for you, too!" Implicit in this messaging was the sentiment, "No, it is not *for* Historic West Las Vegas residents only." Consistent with Guthman's analysis, universalist impulses at VRCG, although well-intentioned, resulted in exclusionary practices, in this case of the unintentional, implicit sort that, nevertheless, led to concrete consequences.[18]

DEMARCATING THE BOUNDARIES OF BELONGING

"Boundary work" refers to the processes through which individuals and groups discursively demark some attributes as favorable and others as less desirable in order to create symbolic boundaries to differentiate identities, spaces, and other categories. Sociologist Thomas Gieryn described ongoing practices within the scientific community of attributing authority to certain scientists, methods, and claims in order to establish rhetorical credibility for them and establishing boundaries between science and "nonscience."[19] Similar demarcation processes have been documented in studies of communities.[20] Communities understand themselves, and are understood by others, as much by their internal segmentation and organization as by their external perimeter.[21] Accordingly, questions about "who we are" are as important as decisions about "who we are not."

When Roz and volunteers make decisions about who is part of the community and who is not, or what criteria should be used to determine who can participate, they are engaging in boundary work. In the following sections, I illustrate how garden participants describe what community means at the garden in terms of who the garden is for and who can participate. Then I examine how they describe actual garden participation and the correspondence between ideal and actual involvement. Third, I identify patterns in respondents' rationales and explanations for both their characterizations of ideal community involvement as well as the quotidian realities of participation. Finally, I close by identifying ways in which intentional boundary work efforts are accompanied by simultaneous unintentional effects.

Explicit Inclusion: Intentions and Outcomes

Any group eventually confronts what James Jasper calls the *extension dilemma.*[22] This strategic dilemma concerns how large a group strives, or is willing, to get and how it manages its growth. Groups have to make strategic choices about how exclusive a group will be in accepting individual members. Can anyone be a member? Or just people who fit a certain profile? Is the size of the group most important or its composition? Decisions about these questions impact the character of the group because as criteria for membership gets looser so too does the degree of unity between members' experiences, values, goals, and cultural practices. That is, the specificity of the group diminishes as it becomes more inclusive. Its focus on the tension between inclusivity and exclusivity makes the extension dilemma a useful tool for analyzing how a community garden group engages in boundary work to define "who belongs" in the community place.

Boundary work concerning issues of belonging should be understood not simply as a practical matter for members of the community garden, but as a process that happens all across a given city that helps make urban life possible and pleasurable. Establishing who belongs in a given community helps city dwellers lead meaningful lives and feel a connection to other people in social environments largely comprised of strangers.[23] Places are important anchors for defining who belongs in a community.[24] The public representations of urban places can be "shared" with outsiders or "strangers," and though these strangers may not necessarily become part of the community, the act of sharing the place with them can make them less "strange."[25]

My interviews with gardeners included a variety of questions about community. First, I asked respondents what community means at the garden and who is part of the community. Second, I asked respondents who, in their experience, is actually participating. Depending on the responses I received, I followed up with probing questions about the racial and ethnic makeup of garden visitors and volunteers and from what part of Las Vegas they know, or believe, participants to come. Third, I followed these two lines of questioning by asking, whatever demographic makeup they had described, why they believe this distribution to be the way that it is.

Who Belongs?

Responses about who could be a member of the community at VRCG could be unanimously described as emphatically inclusive. Most respondents did not include any criteria for membership and instead emphasized the *lack* of requirements. Promoting the garden on Facebook, Roz emphasized the absence of a group for whom the garden is designated saying, "We don't have a group. Everything we do and all that we are and have belong to and are accessible to everyone in our valley." In an article about the garden in the *Las Vegas Review Journal* during the summer of 2012, she reiterated this point saying, "The most rewarding part of this garden is the bringing together of community. It doesn't matter what color or religion or age you are. We all have this common goal of working together." Kyle, a core volunteer who works primarily in the garden, summed up who belongs, "We don't turn, or expect, or think that any particular race, or religion, or anything comes here, we're just wide open to anybody." He wanted to be clear that the garden is not for any groups in particular and that no one will get turned away.

Jimmy explained how he described the garden to people who had never heard of it, "We're *tryin'* to *build* a *community*. We got a garden that's common ground. We're tryin' to build a community." When I asked Jimmy who was welcome to be a part of the community he responded, "Everybody, as long as you don't mean harm to another man and respect another man's spot. Troublebreakers [*sic*] I do not tolerate." Jimmy emphasized that the garden is a place where people are coming together to *build* a community, not necessarily a place *for* an existing community.

Kyle later expanded on his description of who can be part of the community at the garden, explaining, "Well, Las Vegas is a community as far as I'm concerned. So, I don't care where you live, you're welcome here. It's the whole freakin' town." Whereas Jimmy's appeal to inclusivity emphasized that all were welcome to join the effort to build a community on the "common ground" of the garden, Kyle's comment, and others he made in the interview, focused on the garden as a place for learning about gardening for all Las Vegas residents, inclusively considered a community. Although they conceived the garden and the concept of community differently, they both arrived at the same conclusion: anyone and everyone were welcome.

Some respondents not only emphasized the inclusivity of participation criteria but also made a point to describe the residents of the HWLV neighborhood, as was the case in Janice's response to my question of who could be part of the "community" at the garden:

> Anybody and everybody 'cause it's community. Now historically this community was an African American community. It's not like that anymore, this community has become more Hispanic. Yeah. The historical black community is now historically Spanish [*sic*].

Janice was a member of the advisory board and had a long history of working in the HWLV neighborhood. Her response here reflected a point I heard her make many times throughout the years, that the demographics and history of the neighborhood were important, and the garden could serve as a node in the neighborhood community, but that as an organization and place, it should remain open to alliances outside the community. For Janice, the garden could help connect the neighborhood community to resources like grant funding from around the valley and, potentially, nationwide.

Others also made distinctions and connections between the garden's relationship to the HWLV neighborhood and Las Vegas more broadly. Some framed this relationship within the context of the broader "green" culture of Las Vegas. Casey explained:

> I think, it's supposed to be serving the greater Las Vegas area as far as having it the first community garden in Las Vegas so it's serving the whole city of Las Vegas. But then more particularly, ideally, I think the intention was to really get the downtown, local community who is underserved and allowing them to have more access to fresh fruits and vegetables. However, those people, it's a different way of marketing to them. So, you know you have to go through the different layers.

Casey makes an important point about Las Vegas being the first community garden not affiliated with an existing group or organization in a city of approximately two million people.[26] Since most people in Las Vegas still do not have community gardens in, or even near, their own neighborhoods, VRCG broke ground and became *the* place to go if one wanted to get involved with a community garden. Furthermore, in a Sunbelt metropolitan area like Las Vegas, master planned communities and HOAs rigidly restrict land use, making community gardens difficult to cultivate. As the first one to open to the public in the city, VRCG filled a void. Casey's observation captures the tension that arose at the garden between serving the adjacent HWLV and the entire Las Vegas valley.

Mary, a member of the advisory board, believed the garden was *for* the Las Vegas residents living nearest the property, but that all of Las Vegas was invited to help create the place:

> I think there was an inclusive notion of community, meaning broad, right? But I think it was especially intended to serve people for whom they otherwise would not have access to healthy, fresh, nutritious food. Or even if they don't have access to it, that they otherwise wouldn't engage them. And so, it was called Tonopah Community Garden it was, and is, in this area that is kind of like desperate feeling a little bit. My sense was that most immediately it was for the people who lived right around there and for whom this was just a bright and warm and dynamic and nurturing community space that previously hadn't been there. And then extended out from there to involve all of Las Vegas. But that there was a kid of local sense to it as well.

She believed the "inclusive notion of community" was rooted in the HWLV neighborhood and "extended out from there to involve all of Las Vegas" in the project of creating and maintaining "a bright and warm and dynamic and nurturing community space." For Mary, the HWLV neighborhood needed a place like the garden more than other parts of the city. She interpreted the "broad" notion of community at the garden to mean anyone was welcome to participate, but the goals should be to meet the needs of the neighborhood.

As these respondents' comments illustrate, Roz and other volunteers made strategic choices that opted for openness and inclusivity when faced with the extension dilemma. They described both socially and spatially loose perimeters to community emphasizing that all types of people were welcome from anywhere. These definitions reflect a prioritization of the quantity of people, not a particular population or neighborhood, involved and impacted by the garden. As Jasper points out, when a group's membership gets larger and more diverse, it can become more difficult to isolate shared goals, values, and aesthetics, and patterns of action with specificity.[27] Definitions of community and specifications of criteria for participation do not dictate who actually participates at the garden and who it actually reaches.

Who Shows Up?

Understanding how people conceptualize the boundaries of inclusivity and exclusivity is important for analyzing the process of boundary work in community formation. Ideals do not necessarily produce desired outcomes, though. Therefore, I now examine how group members describe actual participation in the community place. Examining how group members describe actual participation in the project and compare it to the ideal is important as it helped identify justifications and rationalizations that are as much a part of the boundary work process as the initial demarcating of boundaries.

Tim described the types of volunteers and visitors that were initially attracted to VRCG and how that has changed over time:

> In the beginning it was people who already lived an alternative lifestyle, because those are the kind of people that look for places like this. People who are vegans, what in my time we would call hippies, the nonconformists on the liberal side. I think as time has gone on more ordinary citizens who've heard about this place have come. We have not penetrated the black community the way we wanted to yet, here.

Tim asserts that initially the garden attracted people who were already interested in gardening, organic produce, urban green space and the various symbolic meanings and values typically associated with urban gardens (see chapter 4 for more discussion of what urban and community gardens represent in broader American culture). This echoes Casey's comments above about how the garden initially filled a void in Las Vegas as the city's first large, public community garden. There was, and continues to be, a segment of the city's population that "look for places like this."

Referring to the surrounding neighborhood, Tim also identified the difficulty of attracting neighborhood residents to the garden, which was a persistent theme in interviewees' descriptions of who actually participates at the garden. When I asked Jimmy, arguably the volunteer who has logged the most hours laboring in the garden out of anyone affiliated with the garden, how many repeat visitors or volunteers come from the neighborhood we had the following exchange:

Jimmy: Uh, zippo.
Tyler: Yeah?
Jimmy: Once a month? Once a month from the neighborhood.
Tyler: It's usually people driving from outside the neighborhood?
Jimmy: Right. Right. Right. But from the neighborhood . . . uh, very few.

Jimmy was emphatic in his observation that nearly all people who participated at the garden came from outside the surrounding neighborhood. The prevalence of commuters does not mean that community is not possible at VRCG. Importantly, urban researchers have demonstrated that social ties—the essence of social capital—are not necessarily spatially bound, and that community in American cities today is less likely to be organized around the neighborhood as it used to be.[28] De-spatialization the concept of community helps us use it to look at the physical places within cities where individuals *elect* to congregate and forge the communal ties that constitute community.

Casey echoed Tim's observations regarding the desire to reach HWLV residents but not yet succeeding when she described who she though *was* coming to the garden:

> I think it was not representative of like the micro–local level of the area where the garden is. I think the cultural makeup is like middle-class white people for the most part helping out at the garden. I think Roz is definitely helping pull more of the African American community into the garden. Locally and not, you know micro locally and the extended Las Vegas area. I think she did a great job of that because she was speaking to African American community groups and leadership groups so she was definitely getting them in. You know, downtown, where that was, I don't know personally who exactly lives there 'cause I don't really hang out there, it was really at the garden. It seemed like it was there but it wasn't necessarily, everyone who was hanging out there didn't necessarily live right around it. So it was hard to see.

Even though Casey knew little about HWLV she could tell that the people participating at the garden were not from the neighborhood. She specified that although she hung out at the garden, she did not have exposure to the surrounding neighborhood, which was similar to most "commuter" volunteers and visitors to the garden. She thought that Roz was addressing the underrepresentation of the neighborhood and black residents and that eventually the neighborhood would join those already participating from other parts of the city.

Black volunteers and visitors paid attention to the racial makeup of those they encountered and worked with at the garden. Randy, a garden volunteer who grew up in HWLV, recounted a routine experience he had after working at the garden:

> Very few African Americans are here comparatively speaking. When I come to the garden and I'm out working it's a rare sight to see an African American walk in the door. We've had several plots that have been rented out to some. I'll go home to my wife and she will sometimes just ask me, "How many black folks are participating over there." 'Cause she's come by a couple times and she's just kind of seen the makeup and I'll say, "Oh just me and Jimmy."

Randy's wife was frustrated that mostly white people were participating at a garden located in HWLV. As Randy explained later in our interview, he was not frustrated that white people were there, just that it was *primarily* white people.

Josh described feeling frustrated when he realized the garden was reaching primarily people from outside the adjacent neighborhood:

> I mean I still think the community is the greater Las Vegas community that the garden serves, you know I think that at times one of the frustrations I had doing different projects or working, you know I'd sit there, I'd be at an event and I would look around and I'd be like, "Crap." No one from *this specific community* is here. Nobody from this neighborhood is here. These are all people from the outside that are coming in and it's kind of like the flavor of the month and it's like, "Oh this is so cute. People are growing food in the middle of the desert. Ha ha ha." I felt that way sometimes and it was a struggle for me because I wanted so much that this really kind of serve the neighborhood and not, and still be for the Las Vegas community, but really be focused on specifically on the community that needed it the most because there wasn't access to fresh fruits and vegetables.

Josh reported having the neighborhood in mind when planning events and programs but attracting primarily people from other parts of the city. His comment about the garden being the "flavor of the month" for some people highlighted his concern that by attracting various individuals and groups from all over the city to events, programs, and volunteer days, there was a lack of commitment involved. People could commute to the garden, have a fun morning of planting or learning to cook a seasonal dish, and drive home, perhaps never returning. He wanted people, local residents and outsiders alike, engaged in the events, programs, and daily activities contributing the creation of a community place that gave back to the surrounding neighborhood.

Just as interviewees agreed that the garden was intended to be an inclusive place where all people could be part of the community, there was a strong consensus that despite being open to all people, black folks in general and especially residents of the surrounding neighborhood were not involved or even showing up. By itself, this gap between the intended population for participation and everyday involvement is interesting. My interviews also probed respondents to reflect on both why inclusivity was so important and why they thought the message of inclusivity was failing to garner the support and participation of those who lived closest to the garden and perhaps stood to benefit most from its existence. These accounts provide rich data for exploring further how boundary work processes unfold since the initial demarking of the symbolic boundaries of inclusion are only part of the process. The discursive tools Roz and volunteers employ to explain, rationalize, and justify their goals, achievements, and failures provide additional insight into how community was pursued at the garden.

Making Sense of Participation

Some potential volunteers and visitors perceived Roz and other participants' desire to build a community based on inclusivity as a strategy for bringing resources to HWLV. For others, diversity was an important element of the kinds of experiences they wanted the garden to provide for participants. Sarah expressed this sentiment after explaining how valuable she had found spending time with some of the black core volunteers at the garden:

> I don't know if you went to Summerlin and promoted that the garden was for the African American community [that you] would get the kind of impact that you've been able to have in educating people in the way that it has, at least for me, when you just say there's this really great place and here it is to promote sustainable food systems, and eating right, and coming together as a community, and having a campfire, and doing this and that. I think that's way more impactful. Just saying come be in a part of a community. It's diverse. And no matter who you are this is for you.

Summerlin is a wealthy suburb of Las Vegas. Sarah questions whether making the garden more explicitly focused on the black residents of HWLV would discourage some potential commuter volunteers and visitors from participating, thus missing out on opportunities for interracial interactions. Though research suggests community gardens are potentially fruitful sites for interracial interactions that could lead to positive changes in racial attitudes,[29] Sarah's concern seems to advocate for prioritizing the emotions of suburban residents over the needs and constraints of HWLV residents, in order to not turn them off the former. Since the garden does not market the garden differently to different populations—it lacks the resources for targeted advertising and promotion—trying to recruit participation from such different populations requires trying to speak generically and organize the garden and its programming in ways that appeal broadly.

At the heart of the push for inclusivity was Roz's belief that no one group of people needed or stood to benefit from a community garden than another. It just so happened the land that was donated to her was on the edge of HWLV. Roz described why the approach to community at the garden foregrounds inclusivity in the way that it does. She was responding to a question I asked her about how she saw the garden's relationship to the neighborhood. Specifically, I wanted her to elaborate on her goal of addressing obesity in America through the garden and some statements she had made in an advisory board meeting about how obesity in America is a racialized issue. She rarely spoke about the garden in racialized terms, so I wanted her to elaborate:

> Well it's kind of been different here because actually there's about three or four states that the numbers aren't skewed and the . . . Nevada is one of them to where the obesity problem is really across the board with blacks, Latinos *and* whites. I think the obesity rate for whites might be like 31% and blacks is 37% you know? So, this is this is kind of a unique place. And so I guess even though we're in an African American community, I guess its allowed me mentally to just really focus on everyone, because everyone is hurting, you know, all races are hurting and so race really hasn't played much of a factor out here, you know. In fact, I have *more* white people here than I do black.

The final statement was not said in frustration, but instead as proof that the garden is not for any particular racial group or neighborhood. Similarly, her distinction between focusing on the African American community in which the garden is located and focusing on everyone implies not only that health issues related to diet impact all races at similar rates, that addressing them can be done uniformly despite racial, ethnic, and class differences.

The desire for complete inclusivity, or appealing to *everyone* in an attempt to create community, runs the risk of making assumptions about what *everyone* needs from, wants, and values in a potential community garden, both in terms of the garden and community aspects. Whose values should serve as the benchmark for "normal"? Universalism refers to an assumption that values held primarily by white, or otherwise privileged individuals, are widely shared and "normal."[30] Universalist assumptions or impulses sometimes emanate from the tendency among privileged people to label shared cultural practices, rooted in "lived but unacknowledged" ideological norms,[31] not as "privileged" but instead as "American" or "normal."[32] When universalist assumptions guide efforts to recruit people to form community, it becomes possible, if not likely, that those for whom the messages, values, and aesthetic ideals do not resonate are assumed in need of education, a change of priorities, and responsible for their own marginalization.[33]

Interviewees had a variety of explanations for why they thought most garden participants came from outside the neighborhood. One theme was rooted in the universalist impulses just described. In some cases, this involved denying HWLV had a food access problem. Tim, a member of the advisory board, reflected on his decades of experiences as a black man in Las Vegas and his knowledge of HWLV:

> Well, first of all I think it is true there aren't enough food markets in this area. But you've got to understand something about black people. Our history in this

> country has always been scarcity. So, we've learned to take a lemon and turn it into lemonade. We have a survivor's instinct. It's almost like we've come to expect that out of the world around us so we find ways around it. The majority of black people here have cars so they have gone to get the food wherever it is. It's only the poorest of the poor who haven't been able to get out and get something else. And there's hardly anybody in this community that doesn't have access to a car. I don't care how poor you are. They've got friends who've got friends who've got cars. People are able to get out and get the food that they need. Now it's a question about what you want. There's still a lot of people that eat foods from the way they grew up and so I don't think it's a question of access to food as a lot of other people do, I think the real problem is training people to eat better.

As described in chapter 2, HWLV struggled to secure a full-service grocery store (one with fresh produce) for many years. Even after the arrival of a Buy Low market, local nonprofit organization Three Square has identified HWLV as the most food insecure area in the Las Vegas valley.[34] Food insecurity takes into consideration residents' income, access to transportation, food access, and food prices. Denial or erasure of these neighborhood specific structural constraints shapes perceptions about how to reach the neighborhood residents as well as how to account for their absence.

Roz also challenged the claim that food access was an issue shaping the everyday realities of HWLV residents:

> You know I don't think access is as much of a problem because everyone has a car. They *drive* wherever they want. They drive to movie theaters. They drive to concerts. They drive everywhere. So short of not having a car, *then* you have a problem of access.

Similar to Tim, Roz suggested access to a car is equivalent to access to fresh food. Later in one of our interviews, she elaborated on her position on the role of food access in garden participation:

> When you really get down to the truth of the matter it's *really* not an access issue, not in Las Vegas anyway, not in this community. It's all about education. I mean I think the garden can be a viable source of food if people really got on that health kick. If they started to have this desire to get back to their roots and chemical free and all that kind of stuff, I mean that's what would have to happen.

She denied structural explanations for diet-related health problems and inequalities and argued the problem is not one of access but education. By extension, she implied that neighborhood nonparticipation at the garden is also explained by lack of education. People would have to make the choice to "get on that health kick" and prioritize chemical-free food before the garden would be of interest to them.

The education or enlightenment explanation came up again in other interviews like this passage from a conversation with Jimmy:

> There's a thing called a third eye, or a spirit, OK? And if it's closed, you're not conscious. You're hooked on the distractions of the world, *or purpose*. I mean, literally on purpose so that you don't have to deal with reality. That's the people we'll never reach . . . because we can't beat Wendy's and Popeye's Chicken, and, you know, the fast food place on the corner, we can't beat that, I mean in their minds. Mentality . . . The only people that come in here are people that are either already conscious or just waking up.

Neighborhood residents would need to wake up, or "open their third eye," according to Jimmy before they saw the garden as a viable source of food and community. I asked Jimmy what the garden could do to help those in the surrounding neighborhood see the garden as, if not as a replacement to cheap fast food, at least another option for food. He replied, "Just by being here." For him, all the garden could do was persist. It was up to neighborhood residents to change.

IMPLICIT EXCLUSION: POTENTIALS FOR SYMBOLICALLY BLOCKING ACCESS

Exclusion does not always operate explicitly, like a physical sign stating "members only" or "must be 21 to enter." More subtle cues in a place can block "symbolic access," implicitly communicating to individuals and groups who belongs and who does not.[35] According to Shephard and Smithsimon, all public places are defined by some degree of exclusion and control, and in community places, exclusion is selective and control over the place, including decisions about who is welcome or a (potential) member of the community, is exercised by the users of the place (as opposed to a private entity or the government).[36] Implicit exclusion is common in community places, where official rules about who is welcome are rarely necessary since outsiders tend to feel uncomfortable, unwelcome, or out of place upon entering.

Importantly, implicit exclusion does not have to be intentional. A number of my respondents questioned whether certain policies, practices, and procedures at the garden might be interpreted as exclusionary to some potential participants. One of the most commonly cited potential sources of implicit exclusion was the price to "adopt" a raised bed gardening plot. Plots can be rented for an annual fee of $500, which can be paid in full at the time of adoption or in quarterly or monthly payments. Compared to community gardens throughout the United States, this price is exponentially higher. For example, a fellow sociologist was astounded when I told him the cost of renting a plot at VRCG. He explained that at the community garden where he conducted research, East New York Farms, renting a plot required an annual fee of $10, but that participants received a $7 refund at the end of the year if they cleaned out their plot. For a more local example, in 2010, the City of Los Angeles, whose Department of Recreation and Parks operates a series of community gardens, proposed an annual rate hike from $25 to $120, due to a change in the city's budget that required the Parks department to start paying its own utility bills. Residents were upset that the rate increase, still less than a quarter of the rate at VRCG, would exclude many existing gardeners from continuing participation.[37] Roz routinely insists the price was calculated to reflect the cost of the water used in the timed drip irrigation systems installed in every plot and the soil mixture the garden buys and amends with compost made on site. Southern Nevada gets very little rainfall so plots' primary source of water is from water the garden pays the City of Las Vegas to supply. The price was set in 2010 when the garden opened and only had a few plots and has stayed the same as the number of plots has increased to over thirty.

Peggy expressed her skepticism about the ability to recruit plot renters at such a price.

> I've connected with a number of people that I sense are probably from not far away, and when you say a plot is $500 they say, "Oh I'd really love to do that." But they don't have $500 to invest. You know? Even if you break it down, well, it's only $30 whatever a month, that's still a lot of money.

Peggy highlights that participation at the garden is not simply a matter of interest. Roz and many volunteers try to rationalize the cost of renting a plot for a year by breaking down the cost into quarterly or monthly payments, but as Peggy points out, no matter how you break it down, the cost of renting a personal space to grow food at VRCG is out of many people's budgets. Randy believes that for some lower-income residents, paying the asking price for renting a bed at the garden may be financial irresponsible:

> I think when you look at the amount for one year it seems exorbitant, but when you look at the break down of it it's reasonable in a way of speaking in terms of your water and your soils and supplies so it may be seen as, by one person that, "I don't wanna'," and "I can't afford to invest in it and just going over to partake of a hobby it may be but it's not going to sustain me throughout the

> year all that I need for my family." So it might be perceived as being a kind of mismanagement of much needed funds. On the other hand, it may be that those who are coming here can say, "Sure, I can afford to participate and do these kinds of things. I can make this contribution." And that may be one of the aspects of why we see certain individuals here from the community and not others.

Randy believes the cost of renting a plot shapes who participates at the garden, at least in terms of who rents plots. He points out that a garden plot does not replace the grocery store. The food grown in one cannot sustain a person, let alone a family, for the year. Accordingly, renting a plot is more of a leisure activity or "hobby," and Randy thinks many people who might be interested in participating decide $500 per year is too expensive of a hobby, especially when they will still need to buy groceries.

George often brought friends to the garden to hang out or show them around. He described many of them as being impressed with the property and being interested in growing their own food there. Once they heard the price, they changed their minds:

> Even a lot of my friends get turned away because of [the cost of renting a bed]. They get turned off because of it. You know, once they hear the price tag they're like, "Hmm, I'll keep buying my stuff from the store." I feel a lot of people get turned off by it. I wanted a bed here, but the price kind of scared me away, too.

George was a core volunteer, but he could not justify paying $500 to rent a plot for the year. He was not sure he could grow enough food to offset the price. Again, he and his friends decided against renting a plot not out of disinterest, but based on an economic cost/benefit analysis. It just did not make financial sense to them. As Randy posited, given the size of the plots and the cost to rent them, participation in the "Adopt-a-plot" program was less of a fast food or grocery store replacement than a large donation fee to be able to engage in a hobby.

The quarterly and monthly payment options were added to the program to make renting a plot less costly upfront. Even if one was able to grow $500 worth of food in the year-long rental plot, many people prefer to pay for their groceries as they consume them. There were no other institutionalized options for plot rentals to help lower-income (or simply frugal) residents participate. George wanted to create an option enabling people to rent sections of plots for reduced prices. As he described, an interaction with some families from HWLV when he was volunteering at one of the harvest festivals led him to the idea:

> There were a handful of people walking around looking a little confused. I walked up and asked how they heard about [the event]. We had sent out some flyers to the local schools and stuff, and actually some of the kids took the flyers home and the parents read and came to check it out. Some of them are working on getting some beds put together. We're going to split 'em amongst a couple families so they could afford it and get some food growing. You know I just thought that was really neat that *finally* we have somebody who lives around here who is going to utilize the garden. I know a lot of people who come here live in all sorts of different ends of the valley. Like I said, I definitely wish more locals would use it.

Residents could also procure food from the garden by purchasing it from the "You-Pick" farmers market. Basically, individuals could go to the garden and pick from the 21 planting rows behind the raised beds. There were usually signs listing what was ready to pick and how much to pay for different quantities of them. Sometimes volunteers were present to assist shoppers, but not always. This was the other option for individuals who wanted to get fresh food but could not or did not want to rent a plot. While this provided an opportunity for residents to pay for just what they picked, which would be significantly cheaper than investing in a plot, they did not have control over what was planted and there was no guarantee the listed produce would still be available when they visited.

The choices about what to plant were made primarily by Roz. These choices raised another issue for some volunteers regarding the implicit exclusion of some Las Vegas residents. There was a sense that a gap existed between what neighborhood residents would be interested in buying and eating. Roz acknowledges that the garden cannot be a one-stop location for all of a person's groceries:

> In order for you to make a garden your, you know, second home you really got to care about [chemical free, local food]. If you don't care about that, you're just going to go to your nearest grocery store 'cause now you can get everything in one spot as opposed to making two and three trips. You know you come here all we got is kale so they still got to go to the store anyways so it has to be something more. It has to just be caring enough to say even if I can only eat one or two things that were fresh off the ground, you know, I do want to eat something off the ground.

Here Roz is echoing her comment above about residents needing to make changes in their lifestyles and commit to buying food from the community garden, even if that means making a special trip for one of the few things available. Moreover, the example she gives to illustrate that often times there are only select items ready for harvest highlights some volunteers' concerns that in addition to limited options for purchase, the crops Roz chooses to plant are not part of many residents' typical diets. Kale is a great example because it illustrates how the food available at the garden is potentially implicitally exclusionary. Kale has become a cliché for many in America who feel it has been pushed down their throats by health-conscious diet advocates. It is perceived by many as one of the most recent health food fads. This is in part because many health-conscious people embrace kale as a sort of superfood. Still others have simply never heard of the bitter leafy green and do not know what to do with it.

The mere presence of kale at the garden is not exclusionary. To be more precise, volunteers like Randy believed the garden did not take into enough consideration the kinds of produce neighborhood residents would want when planting took place. He believed there ought to be a concerted effort in place to ensure people would be able to find foods they cook with frequently, not just seasonal variety:

> [Vegas Roots is] on the fringes of course of the area that I've described as the Historic Black Westside. So, I think the potential is there to advertise and get people to come by if the right kinds of produce are in place. May I say the collard greens, and the okra and those kinds of things and all and people would come and partake of it, but I think most likely there is a need to develop a way to have those things available for people just to purchase as opposed to come and pick your own because many people are reluctant to come out and actually take activity into the garden.

He also speculated that not all residents would be as excited "to eat something off the ground," as Roz emphasized, and that many may be more interested in simply stopping by and buying food without having to harvest it themselves. He expanded on this last point summarizing his belief that the garden needed to meet neighborhood residents where they were instead of trying to get them to change:

> Connect it closer to the community as opposed to saying let's find a way to get people to come over here and harvest their own. I think the attitude is that people are basically going to say I don't want to do harvesting, I'd rather do purchasing and consuming. (Laughs)

Randy did not believe the garden needed to change its entire approach. He did acknowledge, however, that a lack of participation from the surrounding neighborhood did not just mean a lack of interest in fresh food and community. He believed the garden needed to do more to reach into the neighborhood, and to invite the neighborhood and broader black community specifically into the garden by offering symbolic access through what they will find growing in the garden.

THE PARADOX OF INCLUSION

Inclusivity is a principle often intended to "enlarge the tent" and draw in people who may not be aware a group or place is *for* or *open* to them. As the examples provided in this chapter illustrate, however, the manner in which inclusivity is promoted, and the level of importance inclusivity attains in a group's goals or definitions of membership can influence the degree to which the group is truly inclusive. At VRCG descriptions of who the garden was for and who could participate tended to emphasize inclusivity, describing Las Vegas as the "community" from which the community garden drew participants. Some respondents explained that although "community" at the garden tended to mean it was for anyone interested in participating, the focus of the garden was on the HWLV neighborhood. What Roz and some volunteers made clear, though, was that they did not want Las Vegas residents to think the garden was *only* for HWLV residents.

As social and spatial criteria for group membership becomes more inclusive, the degree of specificity that can be used to describe the group and its values diminishes. Over time, the stated mission and vision of the garden became increasingly more general reflecting the broadening of understandings regarding what could be accomplished through the garden and who would benefit from its existence. The shift from initially focusing on the neighborhood and poor Las Vegas residents to one more generally about educating all Las Vegas residents about healthy lifestyles and weight loss was accompanied by the adoption of universalist impulses in both garden actions and the reflections of Roz and other volunteers on the successes and failures at the garden.

Inclusive notions of community mixed with universalist mentalities resulted in an approach to alternative food that emphasized vegetarian and vegan diets, raw food, juice cleanses, and vegetable smoothies. The garden's inclusivity was based on people not simply wanting to grow their own food, but also subscribing to, or at least tolerant of, a particular variety of alternative food discourse predicated on middle- to upper-class values and lifestyles. Roz and others at the garden explained the lack of participation by members of the surrounding neighborhood in terms of a lack of education about the importance of dietary change and foods grown locally without pesticides.

A paradox emerged at the garden whereby attempts at expanding the inclusivity of the garden created additional potentials for the exclusion of those who lived closest to it and stood to benefit most from its presence. Efforts of explicit inclusion were based on not alienating those who lived outside the neighborhood surrounding VRCG by emphasizing the garden was not a neighborhood garden but a public one. These efforts drew the focus of the garden further away from the neighborhood and the values, practices, and policies shifted from food justice focuses to more middle-class alternative food priorities. This shift created increased potentials to implicitly exclude neighborhood residents from participation.

As some volunteers suggested, there are some small adjustments VRCG could make to be more in touch with the surrounding neighborhood. One suggestion for reaching HWLV residents was to focus on having consistent yields of foods neighborhood residents are familiar with and enjoy. Instead of trying to interest residents in community gardening *and* convert them to a diet of many new foods, people like Randy thought VRCG should start with more modest goals and meet residents where they currently are. Additionally, Janice pointed out the historically black neighborhood has shifted demographically and now has a large Latino population. George thought the garden needed to reflect that and have more Spanish speakers present at the garden and engaged in community outreach.

Alternatively, the garden could practice more overt exclusion. Staeheli studied community gardens tended by low-income, racial and ethnic minority communities and found that some of the gardens actively resisted the efforts of wealthy, white, commuter participants.[38] These gardeners were wary that these outsider gardeners might stifle the development of the marginalized gardeners' voices, or shift attention away from the concerns and important issues they identify since the more privileged gardeners do not share the same quotidian struggles. This may not be popular at VRCG, but it calls attention to the utility of deliberate, directed exclusion in community organizing.

The cost of renting a plot is perhaps the most prominent implicitly exclusive aspect of the garden. An annual fee of $500 is not just a little out of most neighborhood residents' budgets; it is an exorbitant price to tend a small plot at the garden. Over the years, many of the plots have been adopted by companies, corporations, and organizations. Sometimes their employees tended the plots, but most of the time, they were more or less neglected by the adopters. These plot rentals were intended to be donations to the garden, and perhaps small attempts at advertising since each plot had a sign indicating who adopted it. These groups could have just as easily sponsored a plot for a family in the neighborhood who could not afford the rental fee. The sponsorship approach I just described was brought up a few times throughout the years but was never formalized into a program so no families came forward requesting financial assistance for a plot and no companies were offered the opportunity to donate one.

If the garden is to truly connect with the neighborhood, there will need to be a concerted effort made to involve residents in the planning of events, of crops grown, and of programs developed, each based on the needs and values of the neighborhood residents. This will require a power redistribution in which a good deal of it is ceded to neighborhood residents and other volunteers. The likelihood of this occurring is questionable (see chapter 5), but to be inclusive requires welcoming not just a diversity of people but also ideas.

NOTES

1. Stephen Carr et al., *Public Space* (Cambridge; New York: Cambridge University Press, 1993).
2. Anthony M. Orum and Zachary P. Neal, *Common Ground?: Readings and Reflections on Public Space* (New York: Routledge, 2009).
3. Benjamin Shepard and Greg Smithsimon, *The Beach Beneath the Streets* (Albany, NY: State University of New York Press, 2011).
4. Carr et al., *Public Space*, 19.
5. Alex Krieger, "Reinventing Public Space," *Architectural Record*, June 1, 1995; Shepard and Smithsimon, *The Beach Beneath the Streets*.
6. Krieger, "Reinventing Public Space."
7. Kurtz, "Differentiating Multiple Meanings of Garden and Community"; Nettle, *Community Gardening as Social Action*.
8. Thomas F. Gieryn, "Boundary-Work and the Demarcation of Science from Non-Science: Strains and Interests in Professional Ideologies of Scientists," *American Sociological Review* 48, no. 6 (1983): 781–95; Michèle Lamont and Virág Molnár, "The Study of Boundaries in the Social Sciences," *Annual Review of Sociology* 28 (2002): 167–95.
9. Jasper, *Getting Your Way*.
10. Zelner, "The Perpetuation of Neighborhood Reputation."
11. Shepard and Smithsimon, *The Beach Beneath the Streets*, 18.
12. Thomas F. Gieryn, "A Space for Place in Sociology," *Annual Review of Sociology* 26 (2000): 463–96.
13. Shepard and Smithsimon, *The Beach Beneath the Streets*.
14. I have replaced their Shepard and Smithsimon's (2011) use of "space" with the more appropriate concept "place" throughout the remainder of the chapter.
15. Shepard and Smithsimon, *The Beach Beneath the Streets*.
16. Mike Davis, *City of Quartz: Excavating the Future in Los Angeles* (New York: Verso, 1990).
17. Guthman, "'If They Only Knew.'"
18. Guthman, "'If They Only Knew.'"
19. Gieryn, "Boundary-Work and the Demarcation of Science from Non-Science."
20. For a summary, see Lamont and Molnár, "The Study of Boundaries in the Social Sciences."

21. Lamont and Molnár, "The Study of Boundaries in the Social Sciences."

22. Jasper, *Getting Your Way.*

23. Daniel J. Monti, *The American City: A Social and Cultural History* (Malden, MA: Blackwell Publishers, 1999).

24. Michael I. Borer, "Important Places and Their Public Faces: Understanding Fenway Park as a Public Symbol," *The Journal of Popular Culture* 39, no. 2 (2006): 205–24.

25. Elijah Anderson, *The Cosmopolitan Canopy: Race and Civility in Everyday Life* (New York, NY: W. W. Norton & Company, 2011); Borer, "Important Places and Their Public Faces."

26. There had been a few community gardens in Las Vegas prior to Vegas Roots opening, but they were attached to senior centers, schools, and churches.

27. Jasper, "A Strategic Approach to Collective Action."

28. Barry Wellman and Barry Leighton, "Networks, Neighborhoods, and Communities Approaches to the Study of the Community Question," *Urban Affairs Review* 14, no. 3 (1979): 363–90.

29. Kimberly J. Shinew, Troy D. Glover, and Diana C. Parry, "Leisure Spaces as Potential Sites for Interracial Interaction: Community Gardens in Urban Areas," *Journal of Leisure Research* 36, no. 3 (2004): 336–55.

30. Guthman, "'If They Only Knew.'"

31. Audrey Kobayashi and Linda Peake, "Racism Out of Place: Thoughts on Whiteness and an Antiracist Geography in the New Millennium," *Annals of the Association of American Geographers* 90, no. 2 (2000): 392.

32. Ruth Frankenberg, *White Women, Race Matters: The Social Construction of Whiteness* (Minneapolis: University of Minnesota Press, 1993).

33. Carolyn Finney, *Black Faces, White Spaces: Reimagining the Relationship of African Americans to the Great Outdoors* (Chapel Hill: The University of North Carolina Press, 2014); Guthman, "'If They Only Knew.'"

34. Three Square, "2018 Food Insecurity Map," *Las Vegas*, NV, 2018, http://www.threesquare.org/learn-about-hunger/research; United States Department of Agriculture Economic Research Service, "Food Security in the U.S.," 2018, https://www.ers.usda.gov/topics/food-nutrition-assistance/food-security-in-the-us/key-statistics-graphics.aspx#foodsecure. Three Square uses data from Feeding America's "Mapping the Meal Gap" research, to identify food insecurity rates by zip code throughout the Las Vegas Metro Area. In 2017, the household food insecurity rate in all of Clark County was 12.8 percent, compared to a national average of 11.1 percent. The food insecurity rate in HWLV was 25.5 percent.

35. Carr et al., *Public Space.*

36. Shepard and Smithsimon, *The Beach Beneath the Streets.*

37. Kate Linthicum, "Plotting a Fee Increase at Community Gardens," *Los Angeles Times*, November 21, 2010, http://articles.latimes.com/2010/nov/21/local/la-me-community-garden-20101121.

38. Lynn A. Staeheli, "Citizenship and the Problem of Community," *Political Geography* 27, no. 1 (2008): 5–21.

Chapter 7

Committing to Community

At the time of this writing, the garden is ten years old and is navigating how to attract people to the garden to volunteer and purchase produce while maintaining safety during the COVID-19 pandemic. Over the past year, its social media presence on Facebook and Instagram has featured the news that two former Democratic presidential candidates, Deval Patrick and Tom Steyer, visited VRCG. Roz was featured in a campaign ad endorsing Tom Steyer leading up to the Nevada caucus in 2020, which was recorded at VRCG. Prior to this news, the focus was on soliciting year-end charitable donations and urging supporters to pick VRCG as the nonprofit organization to support while Black Friday shopping on Amazon.com.[1] The garden has continued the "Lil' Roots Garden Club," which allows kids under the age of thirteen to rent a small raised bed and attend weekly classes on gardening, cooking, and yoga. The garden's e-mail newsletter has recently featured opportunities for supporters to donate money to "sponsor" a family who cannot afford to rent a plot. The price for renting a raised-bed plot dropped from $500 per year to $350. Roz told me that for the past few years the garden's revenue has been over $100,000 per year and that she now has six paid employees.

In the time since I left the field, VRCG has expanded the "Veggie Buck Truck" mobile farmer's market. After operating the program for two years, making stops in food insecure neighborhoods and accepting EBT, the garden accepted a $99,500 grant from Dignity Health to start bringing the truck to sites throughout the Las Vegas suburb of Henderson, NV.[2]

Although my time in the field is over, it seems Roz and her volunteers have experienced another turning point over the past few years, entered into a new sequence, and their actions are no longer simply concerned with the survival of the garden. Thanks in part to multiple grants, they are once again future-oriented.

The garden is now confronting a major change coming to Historic West Las Vegas (HWLV). In the early summer of 2020, the City of Las Vegas began the roll out of "The HUNDRED Plan," a redevelopment initiative in HWLV.[3] During the time of the F Street closure and subsequent reopening (see chapter 2), then-Councilman Ricki Barlow solicited the help of the UNLV Downtown Design Center (DDC) to craft a redevelopment plan for HWLV that would ensure that any of the, what seemed to many inevitable, future investment in the neighborhood, would include the existing residents.[4] In 2016, after a new director took over the DDC, a group of UNLV professors and students in the school of architecture, with other outside consultants, conducted community design charettes and a series of follow-up community and other stakeholder feedback sessions.[5] Their goal was to collaborate with neighborhood residents to create a plan that they would like and support, was technically sound, and would be able to be grafted onto the city's broader long-term development plan for downtown Las Vegas.

The resulting HUNDRED Plan is expansive, including creating retail hubs and affordable housing along historic neighborhood corridors through mixed-used infill, increasing workforce development opportunities and access to green spaces, designing signage and "gateways plazas" to increase the sense of place, and renovating the Moulin Rouge casino property into an entertainment district. City Council eventually voted to add the HUNDRED Plan to the 2035 Downtown Master Plan. Boosters of the plan talk about potentials for investment and tourism to come back to HWLV.[6] Some residents worry about empty promises and displacement.[7]

In addition to the efforts made to include the neighborhood in the planning phase, the implementation plan suggests the creation of a Community Development Corporation (CDC) intended to include neighborhood residents in the shaping of redevelopment actions. Also included in the CDC would be city business leaders and representatives from UNLV, the Urban Land Institute, and the Downtown Project,[8] which are the kinds of constituencies associated with what Logan and Molotch call "the urban growth machine."[9]

Part of the HUNDRED Plan's proposal includes multiple urban agriculture installations. One involves reopening James Gay Park,[10] which was closed and gated in 2013 in response to increased homeless encampments and drug use, and installing greenhouses and harvesting facilities in partnership with the Las Vegas Culinary Academy and the University of Nevada Cooperative Extension.[11] The other is a second campus for the Culinary Academy which is planned to include an "urban agriculture experience." Both sites are envisioned to provide produce to the Culinary Academy, to be managed by the Cooperative Extension, and to include "programming and operations" that involves HWLV residents.[12] Both are roughly two miles from VRCG.

Roz was not consulted in the plans, nor was VRCG mentioned in any of the numerous reports on the plan. The garden was not offered any of the hundreds of millions of dollars that are on the table for these redevelopment efforts either. The greenhouses and farms designed for the redevelopment initiative are designed by professional architects, will be managed by master gardeners from the University of Nevada Cooperative Extension, and located on city property. In nearly every way possible, these will be the opposite of what Roz envisioned. Despite the major differences, there is one critical similarity between VRCG and the organized garden projects that are part of the HUNDRED Plan—both started gardens before a community of gardeners bought in.

Roz is upset about the plans for redeveloping HWLV. She identifies the HUNDRED Plan as gentrification and feels like the proposed changes will force out long-time residents. She is also nervous that the planned new urban agriculture projects' proximity to the garden will impact volunteer and visitor interest in VRCG. From the garden's inception, since the land was donated, there has always been the chance that if property values surrounding the garden increased, the owner could sell to a developer. Since research has demonstrated that introducing community gardens to a neighborhood can increase nearby property values,[13] if VRCG had been *too* successful, it ran the risk of contributing to its own displacement. Until recently, however, the neighborhood had been mostly left behind by the city, neglected and disinvested, especially the area immediately surrounding the garden.

The city's initiation of the HUNDRED Plan represents VRCG's first potential conflict over urban space. For ten years, the garden struggled with paying water bills, attracting neighborhood participation, and volunteer turnover, but not the kinds of conflicts over land use, real estate investment, and redevelopment so commonplace in the literature on urban community gardens.[14] The redevelopment initiative may change that. Perhaps the CDC proposed in the HUNDRED Plan will help the existing residents, as well as the much larger community of previous residents who still feel a sense of ownership over the neighborhood,[15] meaningfully participate in the redevelopment process. The best-case scenario might be a reinvestment in the neighborhood that maintains the character and culture of the community while also empowering existing residents.

On the other hand, the plan could result in what scholars have termed, "green gentrification," a process whereby the alleviating ecological hazards or creating and expanding green space is leveraged to attract wealthier residents and capital investment.[16] Scholars typically conceptualize the displacement associated with this phenomenon as a by-product of city planning efforts that emphasize ecological values and sustainable amenities ostensibly accessible to all members of a community.[17] Even though the HUNDRED

Plan is trying to include neighborhood residents, these sorts of planning initiatives risk creating new vulnerabilities through the promotion of green amenities to spur growth.[18] Indeed, food justice projects like farmers markets and community gardens run the risk of contributing to green gentrification in the places they are created.[19] Joshua Sbicca documented how grassroots urban farms had been encouraged and supported in Denver to revalorize land in the wake of the Great Recession, only to be displaced as property values spiked, property owners refused to renew the farmers' leases, and sold to developers who sought to attract new residents to luxury housing in part through its proximity to the community gardens it was in the process of supplanting.[20] In an interesting parallel to the HUNDRED Plan, the redevelopment efforts in Denver include new urban agriculture projects like vertical farming greenhouses that grow greens for local restaurants and some remaining food sold via a CSA. The potential for this kind of "gentrification of urban agriculture itself,"[21] is part of what scares Roz about the plan in HWLV.

Roz's opposition to the plan stems primarily from her frustration that it has not included her and VRCG. She said, "I wasn't consulted for input, wasn't asked to be a part, and wasn't offered financial support from the hundreds of millions of dollars that will be invested. What a slap in the face for me!" She has reached out to the city to inquire if plans are finalized in hopes of getting the garden included in future plans and getting some of the money flowing into HWLV to help keep VRCG afloat. Community gardens in Seattle have managed to accomplish similar feats of avoiding displacement as redevelopment efforts are initiated in their communities by aligning themselves with the growth coalition, supporting development. Although these gardeners have been successful in protecting gardens, the very populations they argue would benefit most from their presence are increasingly priced out of nearby housing, and the gardens have become symbols of bourgeois lifestyles associated with gentrification.[22] Green gentrification is a dynamic process, in which gardens like VRCG and those proposed by the HUNDRED Plan can help top-down redevelopment efforts gain traction and momentum or function as a model for an alternative vision of urban places and provide a proving ground for community resistance.

It is in this context that the garden's *potential* status as "differential place" has come into focus.[23] These are urban locations where the dominant logic of urban space (e.g., the elevation of exchange value over use value) gets subverted. Roz never approached community gardening as a political action. The specter of gentrification could spark another turning point and initiate a new, more activist-oriented sequence in the garden trajectory.[24] If HWLV community members in opposition to the HUNDRED Plan start looking for places outside the reach of the growth coalitions,[25] where land in or near HWLV is valued for more than its exchange value, the garden may start to stand out more as a "differential place" and the kinds of interactions and potentials there could start to be more attractive to nearby residents.[26]

In other words, the implementation of the HUNDRED Plan may reveal what Eizenberg describes as, "the potential power of spatial practices to recalibrate the imagination, disclose new possibilities of being in the world, and spark new energies that can amount to resilience, reworking, and resistance practices."[27] If people perceive the HUNDRED Plan is, or indeed it is, a "green growth machine," that threatens displacement of both neighborhood residents and existing places like VRCG, those spatial practices of development might bring about a shift in the imaginations of residents, and the garden has the potential to be the kind of differential space where resistance could be forged. There is a history in American cities of the very changes that threaten social ties within neighborhood communities mobilizing some urban neighborhoods to resist them, thereby establishing a bond between residents.[28]

What the current moment represents in the garden's trajectory is a potential political opportunity for the garden to gain support, legitimacy, and resources. The political context in HWLV is dramatically changing. If VRCG manages to be somehow included in the HUNDRED Plan, the space it occupies in the minds of residents, planners, and politicians may grow. Such a strategy may also threaten the ways of life for existing HWLV residents.[29] Contrary to some social movement theory, which suggests mobilization occurs when favorable political opportunities arise and collective actors decide to take advantage, it is often groups' strategic use of the current political context, whatever it may be, to pursue their aims.[30] Even though VRCG is not part of the redevelopment plan, Roz and the volunteers may be able to find a way to use this context to better reach the residents of HWLV. However, what may be most important in determining how the garden navigates these changes may not simply be a matter of resources, political opportunities, or strategies, but instead the group dynamics at the garden.[31]

THE GARDEN AT A TURNING POINT

Throughout this manuscript, I have emphasized primarily things that were not working at the garden. In fact, at various points throughout my research and the writing up of my analysis, I have felt conflicted about the tone of the research. How was a book about a community garden in Las Vegas turning out to be so gloomy? After all, many people at the garden, myself included, really enjoyed spending time there and were passionate about the importance of its survival. And what about the fact that the garden has been in operation for six years? Wasn't I being unreasonably critical of the garden if it had been able to persist this long?

Community gardens are not just collections of plants. They are also groups of people. My growing concern as I spent more time in the field was why so many who were initially excited about their participation at the garden soon disappeared and never returned. I was curious about the rapid rate of turnover among volunteers. Any momentum that was achieved dissipated as volunteers disengaged from the project. Numerous times throughout the three and a half years I was in the field, it seemed as though the garden was on the verge of permanent closure. Then, an unexpected infusion of money or a particularly dedicated volunteer, or sometimes both, kept the garden growing for another season.

My analysis focused on some of the issues contributing to the unstable foundation upon which the garden rests. Many volunteers attempted to put down roots at the garden, but nearly none of them took hold and flourished. A garden that survives on a season-to-season basis, or even year-to-year, relies upon a pool of untapped volunteers to step in to carry the project forward. Thus far the garden has been fortunate in this regard. One of the most frequent comments I heard when doing community outreach for the garden, and one of the most common posts on the garden's Facebook page is some version of, "I've lived in Vegas for years, and I had no idea there was a community garden here!" Roz expressed frustration when she heard comments like these, feeling that the promotion and outreach she and others had done was failing to effectively reach all Las Vegas residents. These comments indicated that there were still untapped reserves of potential volunteers, which was reassuring during periods of volunteer turnover. It is unclear how long this can last. At some point will the word get out to everyone interested in urban agriculture? What happens when more gardens start popping up around the valley, and VRCG is not the only place in town to offer the community garden experience? Perhaps this season the garden will hit its stride, and the momentum of the spring and summer growing seasons will carry it well into the future. For that to happen, however, Roz and the current volunteers must commit to centering the "community" element of the community garden and, to adopt a blackjack metaphor, double down on it.

EVERYDAY WORK AND PLAY AT THE COMMUNITY GARDEN

The United States has experienced a boom in community gardening in the past few decades. Community gardens are not simply popular leisure sites; they are increasingly considered to be antidotes for a host of urban social problems. Planners, activists, and politicians have begun including gardens in proposals for urban revitalization, sustainability, and community development.[32]

There is a growing interdisciplinary body of academic literature—including sociology, environmental psychology, geography, leisure studies, and public health—that describe potentials for improving physical, mental, social, environmental, and community well-being via community gardening. These studies are often used to support claims made by the aforementioned boosters of community gardens in U.S. cities. What has been left underexplored is the everyday, practical accomplishment of these potential outcomes. That is, the ways that groups make the leap from fledgling grassroots organization to community institution. This is not a straightforward process; it requires close inspection, interpretation, and analysis.

One reason the process has not garnered much attention is because a great deal of the literature on community gardens focuses on cities such as New York City, Detroit, Philadelphia, Seattle, San Francisco, and St. Louis. These cities have rich histories of urban agriculture. Whether they are rooted in the moral crusade-based vacant lot gardening movement in the nineteenth century, the "Liberty Garden" movement during World War I, the "Victory Garden" movement of World War II, Great Depression-era "relief gardens," or the 1960s and 1970s era vacant lot gardening resurgence,[33] their legacies of urban gardening have greatly shaped contemporary efforts in these places. For example, sociologist Sharon Zukin has documented how many of the formerly grassroots initiatives have become professionalized and have forged institutionalized relationships with administrators and organizations that help them secure money and other resources.[34]

In cities like Las Vegas, where until 2011, there wasn't even a mention of urban or community gardens in the municipal code, and others without similar histories of urban gardening and clashes between residents, developers, and government officials over land and rights to garden, there has been little or no incentive for city officials to develop or support the kinds of relationships and opportunities that support gardening in places where the phenomenon thrives. Las Vegas cannot reproduce the community gardening models of these other cities because they are based on decades long disputes and traditions. Whereas a garden like VRCG might be able to tap into an existing network of gardens if it were located in a city like Portland, Oregon, or Los Angeles, in Las Vegas, it is up to Roz and volunteers to figure out not only how to connect with residents of the city but also to procure water, electricity, soil, and permits.

More research is needed on the day-to-day interactions, dilemmas, obstacles, and opportunities grassroots community gardens face in trying to move from a fledgling group to a community institution like those featured in so many articles, reports, and proposals that are praised for the myriad benefits they can provide for residents, neighborhoods, cities, and the environment.

FOREGROUNDING MUNDANE AND INCHOATE ASPECTS OF COLLECTIVE ACTION

There is a tendency to privilege the exceptional moments of collective action neglecting the more mundane ones.[35] High-energy moments such as protests, parades, victories, and failures are exciting, but they do not make up the majority of the activities that sustain collective action. Though I participated in and observed large events and experienced times of high energy, for example, the numerous times Roz and the garden were publicly recognized, I gave as much analytic attention to the large periods of time between them. There were Saturday mornings when the garden was buzzing with plot renters, volunteers, children, and the radio blasting 1970s funk music, but there were many more mornings and afternoons during which the garden was completely empty of people. I spent more time at the garden with one or two fellow volunteers or plot renters than with large groups because that was what *most* days were like.

Similarly, sometimes the advisory board meetings were exciting, for example, when plans were coming together for an event like the "Riding to Grow" bike race in 2011 or one of the "Taste and Toast" wine events, but *most* of the meetings consisted of mundane conversations about signage for the "You Pick" farmers market or a potential funding opportunity from the AARP. These more common occasions in collective action group formation are not only underrepresented in scholarly works, but provide a window into "the behind-the-scenes and nitty-gritty activities" that sustain groups or that lead to their dissolution.[36]

Scholars tend to study collective action groups that have "made it."[37] When research focuses on the emergence or mobilization of a particular group, the group under examination tends to be one that eventually picked up momentum. There is similarly a tendency to focus on "successful" groups when the focus of research is on the decline of a collective action group. There is a sampling bias involved in studying primarily groups that were able to establish and sustain, at least for a while, a coherent organizational structure, set of shared goals, and means to achieve those goals. At the same time, we neglect a major component of collective action. In addition to understanding what has worked, we need to understand the processes and dynamics of fledgling and failed groups.

CULTURAL CONVENTIONS AND THE PERFORMANCE OF PLACE

The physical and social construction of places like community gardens involves the transformation of urban space into a meaningful place. Because this process includes appropriating meanings from broader cultural worlds or combining, contesting, or otherwise referring to recognizable meanings, it can also be understood as cultural production. Community gardens are cultural products imbued with meanings by those who produce them and interpreted by those who encounter them.

The transformation of a gravel and construction debris-covered piece of property on the edge of HWLV into the colorful, plant-filled garden it is today involved the adoption of a variety of cultural conventions from the broader community gardening movement in the United States. Whereas Howard Becker emphasized the positive, productive effects of relying on cultural conventions in the production of a cultural product,[38] the case of VRCG illustrates the importance of considering *how* conventions are used. Over relying and selectively relying on cultural conventions can undermine their utility.

In order to move the community garden forward, Roz and the volunteers adopted the goals and missions of other gardens located elsewhere. By selecting these preexisting linguistic and communicative conventions, they did not engage in the crucial process of talking through the specific goals and mission of the garden. Roz's comment about the content of the mission statement that "that's just what you say" sheds light on the ways in which cultural production of a place can become a performance of place. In addition to allowing the organization to skip critical steps in determining the character of the community garden, this decision made it possible for the place and the organization to be misinterpreted by potential volunteers and visitors, since the linguistic and communicative conventions carry meanings from broader cultural worlds like the alternative food, food justice, and urban sustainability movements.

Organizational and administrative conventions provided Roz and volunteers with an approach to the garden that resisted outside groups, organizations, and donors from coopting the place and taking over control. Adopting a "user-created" cultural convention whereby outside experts were devalued in favor of a "figure it out as we go" ethos, Roz described the garden as "volunteer powered." This convention was applied selectively, however, since Roz started the garden by herself without any "users" to help create it. User-created community gardens tend to involve the target "community" in the initial planning phase when members collectively construct the mission and vision of the garden. Failing to adopt the "user-created" convention early on at the garden resulted in the lack of a core or target community of "users" to maintain the garden once plants and weeds started growing. By selectively adopting this convention, Roz had to abandon it again later when she started paying Latino day laborers to periodically clean up the garden.

Using cultural conventions to produce VRCG helped members cultivate a garden, but the overreliance and selective reliance on conventions added additional barriers to an already difficult process of cultivating community at the garden. Using the "production of culture" perspective to examine how social actors appropriate broad cultural trends in local contexts helps illuminate how macroconceptions of culture as repositories of symbols and meaning and micro understandings of culture as the creation of shared understanding through interaction are interrelated and interdependent.

STRATEGIC DILEMMAS AND COLLECTIVE IDENTITY

Deciding whether, and to what extent, to adopt cultural conventions necessarily depends upon the distribution of power in a group of people engaged in cultural production. My interest in the relationship between agency and culture led me to explore the processes through which Roz and volunteers made strategic choices when they confronted strategic dilemmas common to emergent groups.[39] Specifically, my experiences as a gardener and a member of the advisory board raised questions about how the garden dealt with the *organization dilemma* and the *shifting goals dilemma*.

The organization dilemma characterizes the situation most groups face when they decide between concentrating power in the hands of a few or distributing it throughout the group and between establishing a formal organizational structure or leaving things more loosely defined and tentative.[40] Both centralized and decentralized distributions of power have been shown to be effective as collective action tactics. So too have more and less formally organized structures proven to be effective under certain circumstances. What the case of VRCG has shown is that decisions about one of these dimensions affect how the other is experienced. That is, no single organizational structure or model of power distribution is always effective or ineffective. It depends on how people use them and on the contexts in which they use them.

Informal organizational structures tend to be most effective under conditions in which social actors are able to contribute to decisions about the group. When informal structures are combined with centralized power and decision-making authority, the pleasures associated with more informal, grassroots participation dissipate. My experiences and observations suggested that participants at the garden did not find the combination of informal organization and centralized power pleasurable. The informal organization of the garden created a need for volunteers to creatively use their individual expertise to help pursue the goals of the garden. Experiences in which individual expertise was ignored or overridden led participants to disengage from the project.

The lack of a clearly defined distribution of labor and norms of information sharing frustrated some volunteers and visitors because knowledge and information about the garden were concentrated with Roz. This left volunteers sometimes feeling in the dark or out of the loop regarding what was going on at the garden both day-to-day and more long term. The personal frustration felt by volunteers was coupled with the frustration of some visitors and volunteer groups who went to the garden and encountered a loosely organized operation where most volunteers were only remotely sure about how things worked at the garden and unable to answer many questions with certainty.

The shifting goals dilemma describes a scenario in which a group must decide whether to continue using their resources to pursue a current goal or to redirect them in pursuit of a new goal.[41] Volunteers expressed their frustration and confusion, as they sensed that things changed frequently and without explanation at the garden. Composting norms, planting practices, payment processes for the pick-it-yourself farmers market, and other practical matters changed season-to-season and sometimes day-to-day. Additionally, the goals of the garden changed over the years. This frustrated core volunteers whose initial involvement was largely due to their perceptions about what the garden was meant to accomplish. Volunteers who got involved to try to help serve HWLV got burnt out when they felt their energies were being used to promote vegan diets and weight loss. Others grew frustrated that the goals for the garden seemed to accumulate, undercutting the ability for Roz or volunteers to adequately pursue any of them.

The case of VRCG illustrated how strategic choices are interdependent. Choices about the organizational structure and distribution of power shaped members' decisions about the goals to be pursued through the garden. Much of the frustration and confusion volunteers expressed about the changing goals revolved around the lack of input that was solicited or considered in the decisions to change directions. The changing goals and norms at the garden made the informal organizational structure *feel* dysfunctional to volunteers. Experiencing the garden as dysfunctional and disorganized and feeling that their opinions and expertise ultimately did not matter, volunteers started to feel uprooted from the garden. The combination of strategic choices inhibited the formation of a collective identity. Volunteers felt they were part of the garden in the sense that they were contributing to it, but they felt the garden was not theirs since they were not able to guide the direction in which it was going.

EXPLICIT INCLUSIVITY AND SYMBOLIC EXCLUSION IN COMMUNITY BOUNDARY WORK

Establishing who belongs in a given community also entails establishing who does *not* belong. Members must establish boundaries in order to perceive themselves as such. At VRCG, the prioritization of communicating the garden was not explicitly *for* HWLV residents resulted in community outreach efforts that abandoned HWLV residents. Efforts to appeal to residents across Las Vegas resulted in the framing of alternative food in general rather than specific terms and the development of programs and policies that were potentially symbolically exclusive.

In order to recruit volunteers in an inclusive way, Roz and volunteers highlighted benefits of participation and values associated with community gardens that would appeal to diverse audiences. The broadening of the target audience over time diverted attention away from goals and recruitment specific to the surrounding neighborhood. The broad, inclusive notion of who the garden was for allowed for the erasure of the diversity of food issues Las Vegas residents faced. Structural issues of food insecurity and food deserts were ignored in favor of issues of personal choice and education. The "problem" the garden was addressing was unhealthy diets, and the solution was framed as a personal choice to eat a mostly vegan diet.

This message resonated with some people. It reflected a privileged segment of the alternative food movement. Kale salads and raw desserts are not universally appealing, though. Some volunteers questioned whether events that promoted raw and vegan diets and limited selections at the farmers market that featured foods not necessarily staples in most people's diets kept neighborhood residents from participating. They also speculated that the $500 price tag for renting a plot not only dissuaded participation but made those who could not afford it feel like the garden was not *for* them, it was for rich people. These factors did not explicitly restrict people from visiting or volunteering at the garden. They did, however, potentially symbolically exclude segments of the city's population, like the residents who lived nearest the garden, through the manner in which they were *interpreted*. The impression the garden organization explicitly *gave* to the public was that anyone could participate. The impression it *gave off* through its actions were likely different.

CULTURE, AGENCY, AND THE BOUNDARIES OF BELONGING

My observations and analysis of VRCG contributes to the body of research on urban people and places that falls under the umbrella of the urban culturalist perspective.[42] Roz and the volunteers and visitors that garden at Vegas Roots both actively used the garden to (re)construct shared systems of meaning and also used these collective cultural constructions to inject meaning and order into an otherwise largely anonymous, chaotic urban existence. Many of the people I talked to and interacted with at the garden considered VRCG to be an important node for them in the city and a place that helped them identify their relationship to Las Vegas.

A major component of the UCP is the assumption that social actors do not simply receive or transmit culture, but they actively engage in its creation. Those who participated at VRCG collectively created and reproduced meaning, symbols, values, and narratives about the garden, themselves, and Las Vegas. Part of this micro, interactional process of creating culture involved adopting meanings from broader cultural worlds and adapting them in the Las Vegas context. This research illustrates how social actors experience and engage with culture often simultaneously at the macro- and microlevel. People move back and forth between them. They create culture, contributing to the macrocultural "repository," and their actions are shaped by broader cultural meanings.

This research also contributes to our understanding of urban culture by documenting some ways that the meanings, images, symbols, and narratives that people create in and about urban places act back upon the place and those interacting in it. The meanings about Las Vegas, food, and community that emerged through interactions between the garden organization and volunteers shaped *how* the garden took shape. The stories that circulated at the garden about Las Vegas, food access, the promises of alternative food, inequalities, and paths for social change shaped the ways in which Roz and volunteers understood what goals were desirable, the available means to achieve them, and the possibilities for change.

In addition to the contributions this research makes to our understandings of urban culture, it also builds upon the social movement literature on the role of agency in collective action. By analyzing the more mundane aspects of group formation at VRCG, I demonstrated the ways in which agentic choices set the garden down trajectories that shaped the ways members perceived future situations, possibilities, and the garden itself. I also illustrated how even after Roz and volunteers settled into habitual patterns of decision-making and interpretation, it was possible to exercise agency, change courses, and develop new patterns of interpretation.

In the right kind of soil, with water and sunlight, plants will grow nearly anywhere. In many ways gardening is a science. *Community* gardening demands attention to the cultivating of plants *as a community*. Vegas Roots' promotional materials frequently feature tag lines like "Yes, you *can* grow food in the desert!" Emphasizing this alleged horticultural feat casts a shadow over an equally important consideration, how to create a sense of community in a city infamous for the transience of its residents. Thinking about Vegas Roots as a community garden, as a noun, as a thing, draws attention to the plants. Thinking about the possible interactions and activities that can happen at Vegas Roots as community garden*ing*, as a verb, as practices, draws attention to people interacting. Figuring out how to successfully grow tomatoes in the Mojave Desert requires technical knowledge. Growing tomatoes together as a community requires not just the application of technical knowledge together, but also the careful consideration of how to organize and implement that knowledge in ways that reflect the strengths, needs, desires, and values of those who make up the community of gardeners. Reflecting on the main conclusions of this book, Roz maintains, "We grew just enough, with just the right people in each season, and served everyone that was supposed to be served during their time at the garden." There is no doubt this perspective is one of the reasons the garden is still growing ten years later.

NOTES

1. Amazon.com allows shoppers to select from thousands of nonprofit organizations to receive a donation through their "AmazonSmile" program. Shoppers are notified that the organization they select will receive a donation of 0.05 percent of the purchase made on AmazonSmile eligible items. VRCG has participated in similar programs at Whole Foods in Las Vegas.
2. Sandy Lopez, "Mobile Farmers Market Receives Grant to Spread to Henderson," *Las Vegas Review-Journal* (blog), February 15, 2018, https://www.reviewjournal.com/local/henderson/mobile-farmers-market-receives-grant-to-spread-to-henderson/.
3. HUNDRED stands for Historic Urban Neighborhood Redevelopment.
4. Jason Scavone, "Overcoming Controversy on the Historic Westside," University of Nevada, *Las Vegas News Center*, July 1, 2016, http://www.unlv.edu/news/article/big-moves-salvage-redevelopment-efforts.
5. Steven Clarke et al., "The HUNDRED Plan for the Historic Westside Community," May 1, 2016, https://digitalscholarship.unlv.edu/arch_fac_articles/31.
6. Kristen Peterson, "Westside Story: The Legacy and Heart of a Historic Neighborhood," *Las Vegas Weekly*, February 4, 2016, https://lasvegasweekly.com/news/2016/feb/04/westside-story-the-legacy-and-heart-of-a-historic/#/0.
7. Nate Berg, "Wary of Change: As Las Vegas's Historic Westside Faces Change, Residents Ask, Who Benefits?" *Landscape Architecture Magazine*, August 11, 2016, https://landscapearchitecturemagazine.org/2016/08/11/wary-of-change/; Scavone, "Overcoming Controversy on the Historic Westside."
8. Zappos CEO, Tony Hsieh's $350 million redevelopment project around the company's new headquarters near the East Fremont St. district in Downtown (or, old) Las Vegas.

9. John R. Logan and Harvey L. Molotch, *Urban Fortunes: The Political Economy of Place* (Berkeley: University of California Press, 1987).

10. Named for a prominent Historic Westside black businessman who also helped desegregate the Las Vegas Strip.

11. City of Las Vegas, "HUNDRED Plan in Action: Aligning the Implementation Strategy for the Historic Westside," *Las Vegas*, NV, June 30, 2020.

12. City of Las Vegas, "HUNDRED Plan in Action."

13. Ioan Voicu and Vicki Been, "The Effect of Community Gardens on Neighboring Property Values," *Real Estate Economics* 36, no. 2 (2008): 241–83.

14. Karen Schmelzkopf, "Urban Community Gardens as Contested Space," *Geographical Review* 85, no. 3 (1995): 364–81; Miranda J. Martinez, *Power at the Roots: Gentrification, Community Gardens, and the Puerto Ricans of the Lower East Side* (Lanham, MD: Lexington Books, 2010); Sharon Zukin, *Landscapes of Power: From Detroit to Disney World* (Berkeley: University of California Press, 1991); Shepard and Smithsimon, *The Beach Beneath the Streets* (Albnay, NY: State University of New York Press, 2011); Eizenberg, *From the Ground Up.*; Sofya Aptekar, "Visions of Public Space: Reproducing and Resisting Social Hierarchies in a Community Garden," *Sociological Forum* 30, no. 1 (2015): 209–27.

15. Berg, "Wary of Change."

16. Sarah Dooling, "Ecological Gentrification: A Research Agenda Exploring Justice in the City," *International Journal of Urban and Regional Research* 33, no. 3 (2009): 621–39; Kenneth A. Gould and Tammy L. Lewis, *Green Gentrification: Urban Sustainability and the Struggle for Environmental Justice* (New York: Routledge, 2016).

17. Sarah Dooling, "Sustainability Planning, Ecological Gentrificaiton and the Production of Urban Vulnerabilities," in *Cities, Nature and Development: The Politics and Production of Urban Vulnerabilities*, ed. Sarah Dooling and Gregory Simon (Farnham, UK: Taylor & Francis Group, 2012), 101–19, http://ebookcentral.proquest.com/lib/csustan/detail.action?docID=866371.

18. Dooling, "Sustainability Planning, Ecological Gentrificaiton and the Production of Urban Vulnerabilities."

19. Alison Hope Alkon and Josh Cadji, "Sowing Seeds of Displacement: Gentrification and Food Justice in Oakland, CA," *International Journal of Urban and Regional Research* 44, no. 1 (2018): 108–23; Taylor Harris Braswell, "Fresh Food, New Faces: Community Gardening as Ecological Gentrification in St. Louis, Missouri," *Agriculture and Human Values* 35, no. 4 (2018): 809–22.

20. Joshua Sbicca, "The Urban Agriculture Fix: Navigating Development and Displacement in Denver," in *A Recipe for Gentrification: Food, Power, and Resistance in the City*, ed. Alison Hope Alkon, Yuki Kato, and Joshua Sbicca (New York: NYU Press, 2020), 93–110.

21. Sbicca, "The Urban Agriculture Fix."

22. Charlotte Glennie, "Gardens in the Growth Machine: Seattle's P-Patch Program and the Pursuit of Permanent Community Gardens," in *A Recipe for Gentrification: Food, Power, and Resistance in the City*, ed. Alison Hope Alkon, Yuki Kato, and Joshua Sbicca (New York: NYU Press, 2020), 154–75.

23. Henri Lefebvre, *The Urban Revolution* (Minneapolis: University of Minnesota Press, 2003); Eizenberg, *From the Ground Up*.

24. Kathleen M. Blee, *Democracy in the Making: How Activist Groups Form* (New York: Oxford University Press, 2012).

25. Logan and Molotch, *Urban Fortunes*.

26. Eizenberg, *From the Ground Up*.

27. Eizenberg, *From the Ground Up*, 14.

28. Jane Jacobs, *The Death and Life of Great American Cities* (New York: Vintage, 1961).

29. Glennie, "Gardens in the Growth Machine."

30. Anguelovski, *Neighborhood as Refuge*.

31. Blee, *Democracy in the Making*.

32. Campbell, *City of Forests, City of Farms*; Alkon et al., "Introduction: Development, Displacement, and Dining."

33. Paul S. Boyer, *Urban Masses and Moral Order in America, 1820–1920* (Cambridge, MA: Harvard University Press, 1978); Hondagneu-Sotelo, *Paradise Transplanted.*; Lawson, *City Bountiful*.

34. Sharon Zukin, *Naked City: The Death and Life of Authentic Urban Places* (New York: Oxford University Press, 2010).

35. Blee, *Democracy in the Making*.

36. Blee, *Democracy in the Making*.

37. Blee, *Democracy in the Making*.

38. Howard S. Becker, "Art as Collective Action," *American Sociological Review* 39, no. 6 (1974): 767–76.

39. Jasper, "A Strategic Approach to Collective Action"; Jasper, *Getting Your Way*.

40. Jasper, "A Strategic Approach to Collective Action."

41. Jasper, "A Strategic Approach to Collective Action."

42. Michael I. Borer, "The Location of Culture: The Urban Culturalist Perspective," *City & Community* 5, no. 2 (2006): 173–97.

Appendix

Urban *Field* Work

Laboring, Learning, and Lounging with Others in a Community Garden

STARTING WHERE I WAS

I started volunteering at Tonopah Community Garden (the garden's original name—it was on Tonopah Drive) in February 2011. Six months prior to this, I had moved from a gated apartment complex in suburban Henderson, NV, that overlooked an asphalt parking lot and was surrounded by other identical complexes to a cute mid-century rental home in downtown Las Vegas with my sister. The house had a beautiful backyard, and the owner told us with a laugh that she had tried to grow vegetables in the flowerbeds just outside the kitchen but had not been successful. I had never considered growing food before, but for some reason when she pointed out the flowerbeds, I decided I was going to take up gardening while living in the house.

My new year's resolution was to get more involved in the downtown Las Vegas community and to figure out how to get plants to grow in the desert before the spring planting season. A web search for "Las Vegas gardening" returned all sorts of resources, but one result for "Tonopah Community Garden" caught my attention. A little more searching produced a video of Roz the director. Her passion prompted me to e-mail her about volunteering and learning about desert gardening. I visited the garden three days later. I would eventually learn that many of my interviewees followed a similar path to the garden.

Five minutes after I arrived that first day I was put to work. As I was leaving a few hours later, exhausted, sweaty, and dirty, Roz asked if I could come back the following day. I could, and I really wanted to. Before long I was going to the garden multiple days a week and was learning how to plant, monitor, and harvest food properly. I was learning the lingo, and I was learning about organization and garden participants. A few months after getting involved I was invited to a garden planning meeting and left the meeting with the title of volunteer coordinator. I had not even considered the garden as a potential research site. Around the same time, I was talking to a professor in the sociology department at UNLV about the plot we had collectively rented and my increased involvement at the garden when he asked, "So you're going to write your dissertation on the garden, right?" Up to that point volunteering at the garden was a leisure activity for me, nothing more. Being a volunteer, gardener, and advisory board member at VRCG before I was a researcher made me an "indigenous ethnographer" in the field.[1]

Emergent grassroots organizations like community gardens do not have many, if any, truly long-term members. Especially among volunteer-based groups that experience rapid turn-over in participation, "getting in" is sometimes less about fitting particular criteria or possessing certain skills or values than it is about being reliable and willing to put in the work. I "got in" relatively quickly not due to gardening expertise or duration of volunteering, but because of my eagerness to help and the density of my volunteering.

The urban culturalist perspective that informed my theoretical approach to this research is also a sort of methodological orientation.[2] The inductive practice of *starting* with a place and then looking for social phenomena present runs counter to the much more common approach of identifying a social problem or phenomenon and then looking for a place (sometimes the best place, or most interesting, or most extreme) where it is present. This is one reason we may see so many fewer studies of community gardens in places not known for a thriving activist and urban agriculture scene.

MAKING SENSE OF THE DATA: INDUCTIVE TRIANGULATION

The data, analysis, and claims in this dissertation are based on three and a half years of participant observation at Vegas Roots Community Garden, in-depth interviews with people involved with the garden, and qualitative content analysis of documents produced by the garden as well as by others about the garden.

I employed the grounded theory approach to data coding and analysis.[3] Following this multistage process required me to simultaneously analyze my data as I collected it, remaining flexible and allowing the data to help guide and refine my observational gaze and interview questions. I began with initial coding, reading each document line by line, including interview transcripts, field notes, news articles, and promotional materials. During this stage of the coding, I described and minimally interpreted responses, carving the data into smaller component parts. My codes were constructed using gerunds instead of nouns in order to focus on action, process, and sequence, instead of simply "topics."[4] While working through this stage, I constantly compared the codes that emerged and wrote memos to help me start piecing together various related processes.

Next, I began focused coding, which involved sorting the codes identifying those that occurred most commonly and those that best represented the data. I then used this smaller collection of codes to interpret larger segments of the data. The dynamics of qualitative research required me to continuously move between initial coding and focused coding practices as I gathered more data in order to refine my codes to best fit the data.[5] As I progressively collected and interpreted data, I also theoretically sampled interviewees, by seeking informants that would help refine existing codes.[6] Importantly, this was to help refine codes, not formulate a representative sample.

Finally, I engaged in theoretical coding, identifying the relationships between focused codes, thereby constructing theoretical stories grounded in the data. This was the stage in which my three analysis chapters took shape. During this stage, I took the coded chunks of data I initially carved out of my data sources and weaved them back together in order to tell different analytic stories that help make sense of the broader research question. Importantly, I continued data collection in the form of observation and interviews throughout the coding process until I reached theoretical saturation, and there were no gaps in the relationships between the focused codes.

FIELD WORK: THE BODY AND LABORING WITH OTHERS

Observant Participation: Gardening and the Ethnographer's Gaze

I first arrived at the garden with the sole purpose of learning to grow vegetables so I could start a garden in my backyard. Before long I had been recruited to be a weekly volunteer. Members of UNLV's department of sociology, including myself, became corenters of a raised bed plot. Shortly thereafter, I assumed the role of volunteer coordinator for the garden. For approximately one year, I sat on the garden's "advisory board" that helped decide what tasks to pursue in the garden, managed planting cycles, pursued funding opportunities, brainstormed public relations tactics, and many other administrative and organizational considerations. On Saturdays and during major events at the garden between the spring of 2011 and the winter of 2013, I was one of a handful of people who wore green shirts with the garden's logo on the front and the word "GROWER" across the shoulders indicating that if people have questions or concerns they should ask one of us. Although nearly all who participated at the garden were volunteers, those of us with garden shirts were granted heightened levels of authority and credibility.

Over the course of the three and a half years during which I was participating on a regular, weekly basis at the garden, my entrée in two distinct but overlapping groups allowed me to engage in "observant participation."[7] Beyond empathic understanding of the points of view of those in the field, by taking the "participation" component seriously and engaging my body and senses in the same work, and at the same time, with those in the field I was able to treat my body and emotions as a field note.[8] This allowed me to reflect on the changes that happen to my body and mind in order to grasp, via my own experience, the formation and transformation of perceptions and values that emerge through physical and sensory engagement among volunteers in the field.[9]

First, on a weekly basis, I showed up to the garden to pitch in with whatever manual labor needed to be done. Sometimes if I started a project with Jimmy on a Monday, like spreading mulch or transferring seedlings from the greenhouse to the hoop house, I would make plans to come back the following day or later in the week to help him finish the project. I volunteered to help with boring, tedious, and painful work like picking weeds, building up the planting rows with fresh soil, or digging pits through caliche clay for composting. By showing up when I said I would, seeing projects through until they were completed, and in general being willing to get sweaty and dirty in order to further the garden, and not just the plot the sociology department was renting, I earned credibility, legitimacy, and respect from other garden volunteers.

Working, sweating, and getting dehydrated and sunburned alongside other garden volunteers was one way I gained the trust, respect, and friendship of others in the field. Simply hanging out at the garden in the afternoons and evenings was another. At the end of the workday, Jimmy and Kyle would drink beer and smoke cigarettes until it got dark outside. I would often stay at the garden and drink beer with them after a day of working, and sometimes I would drop by in the evening even if I had not been working, just to catch up, and bring some Miller High Life. One afternoon while drinking beer in the shade, Jimmy out of the blue told me, "I just want you to know I think it's cool you come here and *hang out*. I know a lot of white people. They see my dreads . . . I think it's cool you come see how other people live."

My interest in spending evenings drinking and talking to them about the garden, their personal lives, and social and political issues not only solidified my acceptance at the garden, but it helped me grasp the connection to the garden that Jimmy and other volunteers developed as not simply a leisure site for periodic gardening, but a home away from home where they could get away from domestic issues, celebrate holidays, and feel safe. I was able to experience what ownership of the place *felt* like, which later allowed me to get a glimpse of what Jimmy and other volunteers experienced when their decisions were vetoed or their autonomy challenged and they lost their sense of ownership.

I was also willing to help with nongarden tasks. One summer, Jimmy was out of town visiting family in California for a few weeks, and when he returned to Las Vegas his wife had moved out of the apartment she had been living in with him and his kids and turned off all the utilities. She had always taken care of the utilities, and he needed them turned back on but had never done it online. I sat with him one afternoon and set up online accounts for water, electricity, and gas for him and got them all turned back on. Others were willing to help me, too. I bought a couch from Craigslist one summer, and Jimmy offered to take me across the city in his truck to pick it up and take it to my apartment. Another time Roz let me borrow the garden pick-up truck to move some furniture.

At the garden, I was known as a graduate student and a researcher, but I believe I was best known as a novice gardener. I embraced this status of socially acceptable incompetence because it invited people to feel they could teach me a thing or two about gardening and the garden itself. My willingness to adopt the student role, whereby I made it clear I wanted to learn from the members of the field, helped me to establish a nonthreatening demeanor.[10] Additionally, my transparency with others in the garden about how recently I came to develop my limited knowledge of gardening techniques and strategies helped temper what little authority I possessed at the site. Hardly self-deprecating, the goal of this style of self-presentation is to express that I am not an expert on gardening who possesses knowledge unattainable by neophytes. Instead, I sought to convey that I was a co-learner, who first arrived at the garden with little to no gardening experience. The specialized knowledge I possessed at the garden was less about actual gardening and more about the organizational norms of the garden, the combination to the tool shed, and the location of the bathroom. Importantly, this mode of presentation expressed to others that knowledge sharing is a two-way street with me.

This discussion of power and authority in the research site is important because a product of the time I spent in the garden volunteering was an extensive knowledge of the structural and organizational inner workings of the garden, the plans for the future of the garden, the whereabouts of other volunteers, the history of the garden, and the ability to identify most of the plants growing. This knowledge communicates a level of authority to some visitors of the garden. Additionally, over time other volunteers started introducing me to newcomers and visitors as "the professor," "doctor," or "the brains of the operation." In addition to making me uncomfortable because I was not yet a professor or doctor, I felt these sorts of introductions shaped how people came to view me and my position at the garden. I would usually sheepishly correct others and say, "I'm actually just a student," or in reference to the "doctor" designation, "Not yet!" In general, though, I fell back on the strategies described above of making it clear that even if I possessed knowledge and expertise about some things, gardening was not one of them.

The cultivation of embodied, "carnal" knowledge[11] through manual labor may seem more straightforward than through participation in meetings, working on grant proposals together, and developing strategic plans. Since I was also "in" among office volunteers and other members of the advisory board, I was not just observing meetings and daily activities to document the interactions and group dynamics. I was an active participant in meetings. My cultural capital from being a graduate student was most useful in these interactions. Aptitude in contributing to grant proposals, conducting online research and presenting findings to the board, and planning events were all skills that I contributed to the group. Sustained collaboration with others on these tasks forged a mutual respect and friendship between me and other office volunteers and advisory board members beyond simply garden related issues. Similar to my experiences working out in the garden, working in the office also involved taking on the embodied dispositions of the other board members. Together we co-constructed norms of dedication, responsibility, and reliability. As volunteers we had to establish how much time and effort we were willing to dedicate to the process of developing organizational infrastructure at the garden. Through participation in the group, I learned how hard I should push in promoting an idea, when to back off, which questions were off-limits or taboo. No one told me where these lines were; I learned to feel them out myself.

As Erving Goffman pointed out, there is a politics of affiliation in fieldwork, and in most social worlds, there are different "classes," and associating with one can impact one's ability to gain entrée with another.[12] At the garden, I came to notice some tension between garden volunteers and the advisory board or anyone who made decisions *for* the garden who did not also work *in* the garden (see chapter 5 for a more detailed discussion). As an astute observer of the daily activities of the garden, I was able to understand this tension, but as an active participant in the manual labor of the garden, I was able to feel the frustration of taking initiative and shouldering important responsibilities in the garden and then being told by individuals who never work in the garden what needed to be done and how to do it. My association with the advisory board could have sabotaged my affiliations with garden volunteers. My sustained participation in the garden labor and informal hanging out with garden volunteers helped me to maintain my credibility despite also being associated with the office volunteers.

I wrote copious field notes on my observations and interactions both as a garden volunteer and at the advisory board meetings. I carried a small notebook, and later an iPhone, to write brief notes to remind myself later about potentially important details for my field notes. I also took lots of pictures, both to visually represent patterns I was observing and hearing about in interviews and as mnemonic devices to help me write field notes once I returned home. I typed my field notes in Microsoft Word to prepare them for systematic coding.

As a participant in a variety of capacities, I was observing the same place and organization from different angles. Adopting an inductive approach to research means I did not come in looking for certain phenomena or patterns or with a hypothesis to test. I also did not enter the field without purpose. Instead, I entered the field with a broad interest in how social actors appropriate broad cultural trends in a localized context. To get at the microlevel, interactional, everyday mechanisms undergirding this group process, and to help identify the specific dynamics at play at VRCG, my observations in the field were influenced by what Kathleen Blee identifies as five broad dimensions important for understanding group formation and collective action.[13]

First, I paid attention to how decisions were made at the garden focusing on the *sequences* of action and interaction, which made it possible to observe how groups strategize their actions. Second, I observed *failed ideas and plans* that either simply did not work or were abandoned. Third, I focused my attention on the *interactions* that lead up to and followed group actions, noting differences of opinions, power dynamics, and emotional responses. Fourth, I watched for the *cultural practices and processes* through which social actors negotiated meanings, abandoned meanings, interpreted situations, and understood themselves as participants at the garden. Finally, I examined how group members *reflected* on decisions, meanings, and conditions, and how these reflections change as the context changes[14]. These five dimensions did not constrain my ethnographic gaze, but instead helped focus it amid the countless unfolding actions and events I encountered in the field. They are informed by cultural and interpretive sociological traditions.[15]

Naturalistic social scientists tend to take on one of a variety of "membership roles" in their research settings. My role as a participant shifted as the research unfolded. When I started the project, my role was best described as "complete membership," which is characterized by researchers who study scenes of which they are members, sharing the values and goals of the group and taking on responsibilities intended to advance the group.[16] Over time, as I began to understand more about the group's dynamics, as goals shifted, and I came to understand my position in group, my role in the research also shifted to one of "active membership." The change in research roles is minimal, as

I continued to take on roles that would help advance the group, but my own commitment to the goals and values of the group wavered.

This shift in membership role was uncomfortable at first. After my first interview with Roz, I left feeling confused about my prior observations and assumptions about the garden and the group's goals. Her comments about food education being more important than food access and that food access is not a problem in Historic West Las Vegas upset me. I caught myself getting frustrated at other times when it seemed volunteers were more interested in conversations about organic vegan diets reversing cancer than connecting the garden to the neighborhood or other low-income populations. I had to step back and remind myself I was not setting out to depict specific people, but the "action and talk of sets of participants."[17] My goal was not to voice my frustrations, but to *give voice* to the garden's participants.

Interviewing: Reflecting on Emotions, Senses, and Interactions

Participant observation helped "tune my body up" to the research site by going through many similar experiences to those of the people I interviewed.[18] Laboring in the triple-digit heat, attending advisory board meetings, helping out at events, for example, helped me "get in" with garden volunteers and administrative volunteers, develop a "deep familiarity" with the garden as both a place and a group.[19] Being tuned in to the site helped me make decisions about who to sample for interviews, what kinds of questions to ask them, and how to interpret their responses.

I conducted twenty in-depth interviews with individuals who had varying relationships to the garden (see table A1.1). Without a functioning board of directors, Roz oversaw many responsibilities at the garden, and as a result, she helped shed light on ongoing changes in donors, garden programs, and budget issues. Her elevated level of control over the garden and her role as the founder made her the nucleus of the group. In addition to two interviews with Roz, I interviewed office volunteers, garden volunteers, core volunteers, one of the leaders of the permaculture test garden, and garden plot renters.

Interviews were semistructured, and I conducted most of them at the garden. Two were conducted at coffee shops, one at the Gay and Lesbian Center of Las Vegas, another at a "First Friday" event in downtown Las Vegas where the garden had an information booth, and two over the phone with volunteers who were out of the state. On average the in-depth interviews ranged between a half hour and over two hours. All of my respondents consented to having our conversation recorded except one, who had some negative things to say about their experience at the garden and was afraid the contents of the interview could be traced back to the nonprofit with which the respondent was affiliated. I transcribed the interviews manually into Microsoft Word documents using ExpressScribe to prepare them for analysis (see table A.1).

The flexibility of the semistructured interviews allowed my interviewees to play an active role in guiding our conversations by highlighting and exploring the issues they viewed as most relevant and other issues I may have overlooked. The goal was to use interviews to explore my research question through respondents' experiences and accounts, not to simply verify my observations or others' responses.[20] The interview schedule included questions about interviewees' prior exposure to gardening and agriculture, their experiences at VRCG, their reflections on the organization and structure of VRCG, their knowledge and perceptions about the Historic Westside, their assessments of what community means at VRCG, the factors that motivated their participation, and changes that they would like to see at VRCG.

I also conducted hundreds of informal interviews with new or sporadic garden volunteers, visitors, and involuntary laborers. These lasted anywhere from one sentence to long conversations. Typically, they started out with me introducing myself and asking them if this was their first time at the garden. If it was, I asked how they heard about the garden and, based on their responses, I probed further, inquiring about motivations for participation, prior experience gardening, perception of VRCG, experience there so far, and where they lived and what they did for work. These often unfolded as I worked alongside respondents or sat with them in the shade next to the garage.

By interviewing volunteers at different levels, plot owners, and the director, I gathered data from a variety of perspectives, capturing the breadth and depth of the ways individuals experienced the garden. Additionally, my interview questions sought to elicit rich, descriptive accounts of respondents' experiences. Since the self is a product of ongoing construction through narrative,[21] and since my ultimate goal is to use the data to tell a story about city life, nature, and community at the community garden, getting my respondents to tell me rich stories provided data that helped me explore the experiences of those involved with the garden.

Content Analysis: How the Garden Is Represented

While engaged in participant observation and interviewing volunteers, I also conducted a content analysis of documents and other textual artifacts produced *by* the garden as well as those produced *about* the garden by outsiders. The data sources included the garden's website, marketing and publicity materials, and signage around the physical garden space as well as textual and visual media representations of the garden from local sources like the *Las Vegas Review-Journal*, the *Las Vegas Sun*, and the *Las Vegas Weekly* as well as national sources like the online environmental news magazine *Grist*.

Table A.1 List of Semistructures In-Depth Interviewees

Name	*Race/ Ethnicity*	*Role at the garden*	*Specific contributions to the garden*
Rosalind "Roz" *+	Black	Founder Director	Makes all executive decisions at the garden. The face and voice of the garden
Sarah	White	Advisory board Designer	Instrumental in changing city code to include urban garden provisions. Helped design the land use plan. Grant writing.
Tim	Black	Advisory board	Worked on the garden budget. Grant writing. Community outreach.
Jimmy	Black	Head gardener	Oversight of planting, tree watering, construction of structures (shade, sheds, plots, signs. Community outreach.
Kyle	White	Irrigation coordinator Roz's K&K employee	Installation and maintenance of drip irrigation throughout the garden. Community outreach. Volunteer coordination.
Randy	Black	Garden volunteer	Implemented and managed multistage composting system. Science and astronomy education at garden events.
Peggy	White	Advisory board	Helped coordinate marketing campaigns for the garden. Plot owner. Community outreach.
Casey	White	Advisory board	Established social media presence for the garden. Plot owner.
Debby	White	Plot renter	Socialized with other plot owners and volunteers. Coordinated drum circles at the garden.
Wayne	White	Plot renter	Socialized with other plot owners and volunteers. Volunteered with planting and worm habitat maintenance.
Josh	White	Advisory board	Managed volunteers. Organized educational programs for the community at the garden. Promoted the garden in the community.
Mary	Latino	Advisory board	Helped the advisory board develop short-term and long-term goals and priorities.
Janice	Black	Advisory board	Worked on garden communications, fund raising, and copy-editing budgets and grant applications.
Rita	Asian American	Plot owner	Socialized with other plot owners.
Angie	White	Plot owner	Socialized with other plot owners.
Lisa×	White	Advisory board	Helped organize events. Worked on garden organization infrastructure.
Justin	White	Permaculture expert	Cocreated the permaculture demonstration garden within VRCG.
Alex	Latino	Garden volunteer Community service	Managed volunteers. Helped with planting and harvesting. Completed court mandated community service in the garden.
George	Latino	Garden volunteer	Helped with planting and harvesting. Assisted with event planning.

* Interviewed twice
+ Used real name because they are a public figure
× Refused to be interviewed on tape.

Initially, I combed the Internet for media coverage that predated my participation at the garden. After getting up to speed I set a Google alert for "Tonopah Community Garden," and then for "Vegas Roots Community Garden" when Roz changed the name. If anything had been posted online about the garden, I would get an alert at the end of the week. Typically, though, I would be looking for articles before they were published because I had either seen, talked to, or heard about a journalist visiting the garden. Collectively, these data helped me see the garden organization's self-perceptions and self-representations as well as representations of it by others.

This component of data collection and analysis provided important data for exploring how the garden group wanted audiences to see the garden, how it wanted members to interpret their involvement, and how others interpreted the place and organization.[22] Visual and text media representations helped guide my observations and interviews by providing necessary context to the experiences, encounters, and interactions I saw and interviewees described. Many of the core volunteers I interviewed reported having first heard about the garden in a radio broadcast or news article. To conduct research on a place without being aware of its presence and persona in the media is to miss an important piece of the puzzle. These data also provided me with an opportunity to observe differences and disjunctures between public representations of the garden and quotidian realities.

POLITICS, ETHICS, AND ETHNOGRAPHY

Ethnographic research presents researchers with myriad ethical and political considerations. One of the central dilemmas of ethnographic research involves validity, and the extent to which any scholarly account can accurately and objectively represent the social world. I approached this research with the understanding that my observations, interpretations, and representations of the phenomena I studied would be partial. Regarding the distinction between "social realism" and more postmodern understandings of ethnographic fieldwork that abandon validity entirely, I adopted a "subtle realist" position and understand that although my interpretations of observations and interviews are not exact or perfect representations of the social world under examination, I sought to garner a "close approximation of the empirical world."[23] As Howard Becker argued, the quality of ethnographic research and reporting is not contingent upon replicability or reproducibility.[24]

In part, then, my task was to provide an account of the ways in which I came to my conclusions and to proceed in a way that is trustworthy to the reader. That is, I embraced Sherri Grasmuck's notion of "objectivity as fairness"[25] and strove to represent all competing perspectives even-handedly and to provide a clear account of how I interpreted the data.

Additionally, the interview process poses representational issues, and I follow Holstein and Gubrium who argue that the interview process is a social production, a meaning-making process in which the interviewer and interviewee are co-constructors of knowledge.[26] This involves not simply adopting a symmetrical relationship with the interviewee in which I take his or her word as truth, but an "asymmetrical reciprocity" in which I reflexively seek to coconstruct the reality of the informant while simultaneously scrutinizing that reality against observed structural conditions and sociological literature.[27] The context of this project, a low-income, largely black neighborhood and many commuting middle- or upper-class suburbanite gardeners cannot be overlooked. Although I have the final say in the manuscript, returning to my codes to refine them and compare them to new data allowed me many opportunities to practice reflexivity and reflect on the power I wield in interpreting my informants' words.

I also made the choice to selectively employ pseudonyms in the presentation of my research. My decisions regarding when to use pseudonyms were strategic, opting only to forego employing them when referring to public places and public figures. All interviewees signed informed consent forms that guaranteed their confidentiality. Roz verbally agreed to allow me to use her real name throughout the research. Since she had no reservations about the use of her name, I wanted to use it. Recently, urban ethnographers have begun to challenge the traditionally unquestioned default method of data reporting that uses pseudonyms for all humans and places.[28] These challenges claim the confidentiality and anonymity promised by pseudonyms is overstated. The tradition of pseudonyms for places was not just an attempt to protect the anonymity of research participants but was also a device to make claims about generalizability.

This latter rationale for pseudonyms is especially troubling to urban sociologists like myself who believe that place matters. Las Vegas is a specific city with a unique history. So, too, is Historic West Las Vegas a specific neighborhood with a specific history. To try to obscure the identifying details and histories of these places obscures reality and obfuscates exploration into the place-based, contextual factors that influence social life. Accordingly, I identify the real names of the city, neighborhoods, streets, and organizations within this manuscript. I also used the real names of other public figures like the property owner's name and local politicians.

After completing a draft of the book, I shared the main findings and critiques with Roz. She provided feedback, most of which confirmed my observations, and the parts that did not have been incorporated into the manuscript (mostly in chapter 7). I also offered to provide her with a pseudonym after reviewing the book, but she confirmed she was comfortable with her real name being used.

WORKING. UNDERSTANDING

The existing literature on urban and community gardens is rich and provides valuable insight into the potential outcomes of participation in and proximity to these sorts of projects. I set out to contribute to this literature by examining the mundane, everyday goings on and interactions that serve as the foundation of grassroots collective action. To be in a position to see and experience these aspects of VRCG, I needed a "right to be close to" everyday volunteers working at the garden.[29] Goffman's cynical view of humanity and social research led him to suggest this "right" to proximity in field research can only be attained "by one sneaky means or another."[30] I just happened to be already spending a great deal of my free time volunteering at VRCG and decided to reenter the field, to which I had already gained entrée, as an ethnographer.[31]

Taking on the role of observant participant allowed me to experience the garden from within, alongside other dedicated volunteers. Being a native ethnographer allowed me to both develop an empathic understanding of other volunteers' experiences and reflect upon my own in order to analyze how mundane aspects of collective action influenced the character of the place and organization.

NOTES

1. Matthew Desmond, *On the Fireline: Living and Dying with Wildland Firefighters* (Chicago: University of Chicago Press, 2008).
2. Michael I. Borer, "The Location of Culture: The Urban Culturalist Perspective," *City & Community* 5, no. 2 (2006): 173–97.
3. Kathy Charmaz, *Constructing Grounded Theory: A Practical Guide through Qualitative Analysis* (Thousand Oaks, CA: SAGE Publications Ltd, 2006).
4. Charmaz, *Constructing Grounded Theory*.
5. Charmaz, *Constructing Grounded Theory*.
6. Charmaz, *Constructing Grounded Theory*.
7. Ashley Mears, "Seeing Culture through the Eye of the Beholder: Four Methods in Pursuit of Taste," *Theory and Society* 43, no. 3–4 (2014): 291–309; Loïc Wacquant, *Body & Soul: Notebooks of an Apprentice Boxer* (Oxford; New York: Oxford University Press, 2006).
8. Desmond, *On the Fireline*.
9. Mears, "Seeing Culture through the Eye of the Beholder."
10. John Lofland et al., *Analyzing Social Settings: A Guide to Qualitative Observation and Analysis*, 4 edition (Belmont, CA: Wadsworth Publishing, 2005).
11. Loïc Wacquant, "Habitus as Topic and Tool: Reflections on Becoming a Prizefighter," *Qualitative Research in Psychology* 8, no. 1 (2011): 81–92.

12. Erving Goffman and Lyn H. Lofland, "On Fieldwork," *Journal of Contemporary Ethnography* 18, no. 2 (1989): 123.

13. Kathleen M. Blee, *Democracy in the Making: How Activist Groups Form* (Oxford: Oxford University Press, 2012).

14. Blee, *Democracy in the Making*.

15. Herbert Blumer, *Symbolic Interactionism: Perspective and Method* (University of California Press, 1969); Borer, "The Location of Culture"; Brian Fay, *Social Theory and Political Practice* (London: Routledge, 1975); Lyn H. Lofland, "Community and Urban Life," in *Handbook of Symbolic Interactionism*, ed. Larry T. Reynolds and Nancy J. Herman-Kinney (Walnut Creek, CA: Rowman Altamira, 2003), 937–74; David A. Snow, "Extending and Broadening Blumer's Conceptualization of Symbolic Interactionism," *Symbolic Interaction* 24, no. 3 (2001): 367–77.

16. Patricia A. Adler and Peter Adler, "Observational Techniques," in *Collecting and Interpreting Qualitative Materials*, ed. Norman K. Denzin and Yvonna S. Lincoln (SAGE Publications Ltd, 1998), 79–109.

17. Gary Alan Fine, "Towards a Peopled Ethnography Developing Theory from Group Life," *Ethnography* 4, no. 1 (2003): 41–60.

18. Goffman and Lofland, "On Fieldwork."

19. Goffman and Lofland, "On Fieldwork."

20. John M. Johnson, "In-Depth Interviewing," in *Handbook of Interview Research: Context and Method*, ed. Jaber F. Gubrium and James A. Holstein (Thousand Oaks, CA: SAGE Publications Ltd, 2002), 103–19.

21. James A. Holstein and Jaber F. Gubrium, *The Self We Live By: Narrative Identity in a Postmodern World*, 1 edition (New York: Oxford University Press, 1999).

22. Blee, *Democracy in the Making*.

23. David A. Snow and Calvin Morrill, "Reflections on Anthropology's Ethnographic Crisis of Faith," *Contemporary Sociology* 22, no. 1 (1993): 8–11.

24. Howard S. Becker, *Sociological Work: Method and Substance* (New Brunswick, NJ: Transaction Publishers, 1976).

25. Sherri Grasmuck, *Protecting Home: Class, Race, and Masculinity in Boys' Baseball* (New Brunswick, NJ: Rutgers University Press, 2005).

26. James A. Holstein and Jaber F. Gubrium, *The Active Interview* (Thousand Oaks, CA: SAGE Publications Ltd, 1995).

27. Rosalind Edwards and Melanie Mauthner, "Ethics and Feminist Research: Theory and Practice," in *Ethics in Qualitative Research*, ed. Melanie Mauthner, Maxine Birch, Julie Jessop, and Tina Miller (London: SAGE Publications Ltd, 2002), 15–31.

28. Michael I. Borer, "From Collective Memory to Collective Imagination: Time, Place, and Urban Redevelopment," *Symbolic Interaction* 33, no. 1 (2010): 96–114.

29. Goffman and Lofland, "On Fieldwork."

30. Goffman and Lofland, "On Fieldwork."

31. Desmond, *On the Fireline*.

Bibliography

Abi-Nader, Jeanette, Kendall Dunnigan, and Kristen Markley. *Growing Communities Curriculum: Community Building and Organizational Development Through Community Gardening*. College Park, GA: American Community Gardening Association, 2001. https://www.communitygarden.org/product-page/growing-communities-how-to-build-community-through-community-gardening.

Adler, Patricia A., and Peter Adler. "Observational Techniques." In *Collecting and Interpreting Qualitative Materials*, edited by Norman K. Denzin and Yvonna S. Lincoln, 79–109. Thousand Oaks: SAGE Publications, 1998.

Alaimo, Katherine, Thomas M. Reischl, and Julie Ober Allen. "Community Gardening, Neighborhood Meetings, and Social Capital." *Journal of Community Psychology* 38, no. 4 (2010): 497–514.

Alkon, Alison Hope, and Josh Cadji. "Sowing Seeds of Displacement: Gentrification and Food Justice in Oakland, CA." *International Journal of Urban and Regional Research* 44, no. 1 (2018): 108–123.

Alkon, Alison Hope, and Julian Agyeman, eds. *Cultivating Food Justice: Race, Class, and Sustainability*. Cambridge, MA: The MIT Press, 2011.

Alkon, Alison Hope, Yuki Kato, and Joshua Sbicca. "Introduction: Development, Displacement, and Dining." In *A Recipe for Gentrification: Food, Power, and Resistance in the City*, edited by Alison Hope Alkon, Yuki Kato, and Joshua Sbicca. New York: NYU Press, 2020.

American Community Garden Association. "Find-A-Garden." ACGA, 2019. https://www.communitygarden.org.

Anderson, Elijah. *The Cosmopolitan Canopy: Race and Civility in Everyday Life*. New York, NY: W. W. Norton & Company, 2011.

Anguelovski, Isabelle. *Neighborhood as Refuge: Community Reconstruction, Place Remaking, and Environmental Justice in the City*. Cambridge, MA: MIT Press, 2014. http://ebookcentral.proquest.com/lib/utsa/detail.action?docID=3339756.

Aptekar, Sofya. "Visions of Public Space: Reproducing and Resisting Social Hierarchies in a Community Garden." *Sociological Forum* 30, no. 1 (2015): 209–227.

Armstrong, Donna. "A Survey of Community Gardens in Upstate New York: Implications for Health Promotion and Community Development." *Health & Place* 6, no. 4 (2000): 319–327.

Banet-Weiser, Sarah. *Authentic(TM): The Politics of Ambivalence in a Brand Culture*. New York: NYU Press, 2012.

Bartkowski, John P., and W. Scott Swearingen. "God Meets Gaia in Austin, Texas: A Case Study of Environmentalism as Implicit Religion." *Review of Religious Research* 38, no. 4 (1997): 308–324.

Barton, Bernadette, and Constance L. Hardesty. "Spirituality and Stripping: Exotic Dancers Narrate the Body Ekstasis." *Symbolic Interaction* 33, no. 2 (2010): 280–296.

Becker, Howard S. "Art as Collective Action." *American Sociological Review* 39, no. 6 (1974): 767–776.

———. *Sociological Work: Method and Substance*. New Brunswick, NJ: Transaction Publishers, 1976.

———. *Art Worlds*. Berkeley: University of California Press, 1984.

Bellah, Robert N., Richard Madsen, William M. Sullivan, Ann Swidler, and Steven M. Tipton. *Habits of the Heart: Individualism and Commitment in American Life*. New York: Perennial, 1985.

Berg, Nate. "Wary of Change: As Las Vegas's Historic Westside Faces Change, Residents Ask, Who Benefits?" *Landscape Architecture Magazine*, August 11, 2016. https://landscapearchitecturemagazine.org/2016/08/11/wary-of-change/.

Berman, Marc G., Ethan Kross, Katherine M. Krpan, Mary K. Askren, Aleah Burson, Patricia J. Deldin, Stephen Kaplan, Lindsey Sherdell, Ian H. Gotlib, and John Jonides. "Interacting with Nature Improves Cognition and Affect for Individuals with Depression." *Journal of Affective Disorders* 140, no. 3 (2012): 300–305.

Berman, Marc G., Michael C. Hout, Omid Kardan, MaryCarol R. Hunter, Grigori Yourganov, John M. Henderson, Taylor Hanayik, Hossein Karimi, and John Jonides. "The Perception of Naturalness Correlates with Low-Level Visual Features of Environmental Scenes." *PLoS One* 9, no. 12 (2014): e114572.

Blakely, Edward James, and Mary Gail Snyder. *Fortress America: Gated Communities in the United States*. Washington, DC: Brookings Institution Press, 1997.

Blee, Kathleen M. *Democracy in the Making: How Activist Groups Form*. New York: Oxford University Press, 2012.

Blumer, Herbert. "Race Prejudice as a Sense of Group Position." *The Pacific Sociological Review* 1, no. 1 (1958): 3–7.

———. *Symbolic Interactionism: Perspective and Method*. Berkeley, CA: University of California Press, 1969.

Bonanno, Alessandro. "Food Deserts: Demand, Supply, and Economic Theory." *Choices: The Magazine of Food, Farm & Resource Issues* 27, no. 3 (2012): 1–4.

Bonilla-Silva, Eduardo. *Racism Without Racists: Color-Blind Racism and the Persistence of Racial Inequality in the United States*. Lanham, MD: Rowman & Littlefield Publishers, 2003.

Borer, Michael I. "Important Places and Their Public Faces: Understanding Fenway Park as a Public Symbol." *The Journal of Popular Culture* 39, no. 2 (2006a): 205–224.

———. "The Location of Culture: The Urban Culturalist Perspective." *City & Community* 5, no. 2 (2006b): 173–197.

———. *Faithful to Fenway: Believing in Boston, Baseball, and America's Most Beloved Ballpark*. New York: NYU Press, 2008.

———. "From Collective Memory to Collective Imagination: Time, Place, and Urban Redevelopment." *Symbolic Interaction* 33, no. 1 (2010): 96–114.

Boston Urban Gardeners. *A Handbook of Community Gardening*. New York: Charles Scribner's Sons, 1982.

Boyer, Paul S. *Urban Masses and Moral Order in America, 1820–1920*. Cambridge, MA: Harvard University Press, 1978.

Braswell, Taylor Harris. "Fresh Food, New Faces: Community Gardening as Ecological Gentrification in St. Louis, Missouri." *Agriculture and Human Values* 35, no. 4 (2018): 809–822.

Breitbart, Myrna Margulies, and Pamela Worden. "Creating a Sense of Purpose: Public Art and Boston's Orange Line." *Places: Forum of Design for the Public Realm* 9 (Summer 1994): 80–86.

Brown, Kate H., and Andrew L. Jameton. "Public Health Implications of Urban Agriculture." *Journal of Public Health Policy* 21, no. 1 (2000): 20–39.

Buechler, Steven M. "New Social Movement Theories." *Sociological Quarterly* 36, no. 3 (June 1, 1995): 441–464.

———. *Social Movements in Advanced Capitalism: The Political Economy and Cultural Construction of Social Activism*. New York: Oxford University Press, 1999.

Campbell, Lindsay K. *City of Forests, City of Farms: Sustainability Planning for New York City's Nature*. Ithaca, NY: Cornell University Press, 2017.

Carr, Stephen, Mark Francis, Leanne G. Rivlin, and Andrew M. Stone. *Public Space*. Cambridge; New York: Cambridge University Press, 1993.

Charmaz, Kathy. *Constructing Grounded Theory: A Practical Guide Through Qualitative Analysis*. Thousand Oaks, CA: SAGE Publications Ltd, 2006.

Choate, Alan. "Community Garden Faces Hitch in Its Water Supply." *Las Vegas Review-Journal*, June 27, 2011. http://www.reviewjournal.com/news/las-vegas/community-garden-faces-hitch-its-water-supply.

City of Las Vegas. "HUNDRED Plan in Action: Aligning the Implementation Strategy for the Historic Westside." Las Vegas, NV, June 30, 2020.

Clair, Robin P. "Narratives in the Old Neighborhood: An Ethnographic Study of an Urban Neighborhood's Stories." *Qualitative Inquiry* 12, no. 6 (December 1, 2006): 1244–1261.

Clarke, Steven, David Baird, Claytee White, and Glenn Nowak. "The HUNDRED Plan for the Historic Westside Community." May 1, 2016. https://digitalscholarship.unlv.edu/arch_fac_articles/31.

Coe, Mary. *Growing with Community Gardening*. Taftsville, VT: Countryman Press, 1978.

Coolican, J. Patrick. "We're Stuck, So Will We Bond? - Las Vegas Sun News." *Las Vegas Sun*, December 25, 2010. http://lasvegassun.com/news/2010/dec/25/were-stuck-so-will-we-bond/.

Crane, Diana. *The Production of Culture: Media and the Urban Arts*. Newbury Park, CA: SAGE Publications Inc, 1992.

Cronon, William. "The Trouble with Wilderness; or, Getting Back to the Wrong Nature." In *Uncommon Ground: Rethinking the Human Place in Nature*, edited by William Cronon, 69–90. New York: W. W. Norton & Company, 1995.

Curtis, Lynnette. "Faith Makes Fruits, Vegetables Sprout on Five Acres." *Las Vegas Review-Journal*, July 19, 2010. http://www.reviewjournal.com/news/faith-makes-fruits-vegetables-sprout-five-acres.

Davis, Joseph E., ed. *Stories of Change: Narrative and Social Movements*. Albany: SUNY Press, 2002.

Davis, Kingsley. 1955. "The Origins and Growth of Urbanization in the World." *American Journal of Sociology* 60, no. 5: 429–437.

Davis, Mike. *City of Quartz: Excavating the Future in Los Angeles*. New York: Verso, 1990.

———. *Dead Cities: And Other Tales*. New York: The New Press, 2003.

Day, Jo. "Plants, Prayer, and Power: The Story of the First Mediterranean Gardens." In *Gardening: Philosophy for Everyone-Cultivating Wisdom*, edited by Dan O'Brien, 65–78. Malden, MA: Wiley-Blackwell, 2010.

Desmond, Matthew. *On the Fireline: Living and Dying with Wildland Firefighters*. Chicago: University of Chicago Press, 2008.

Dickens, David. "Is Las Vegas a 'Real City'?" *UNLV Gaming Research & Review Journal* 15, no. 2 (2012). http://digitalscholarship.unlv.edu/grrj/vol15/iss2/11.

Dooling, Sarah. "Ecological Gentrification: A Research Agenda Exploring Justice in the City." *International Journal of Urban and Regional Research* 33, no. 3 (2009): 621–639.

Dooling, Sarah. "Sustainability Planning, Ecological Gentrification and the Production of Urban Vulnerabilities." In *Cities, Nature and Development: The Politics and Production of Urban Vulnerabilities*, edited by Sarah Dooling and Gregory Simon, 101–19. Farnham, UK: Taylor & Francis Group, 2012.

Dupuis, E. Melanie. *Nature's Perfect Food*. New York: NYU Press, 2001.

Durkheim, Emile. *The Elementary Forms of Religious Life*. New York: Oxford University Press, 1912.

Edwards, Rosalind, and Melanie Mauthner. "Ethics and Feminist Research: Theory and Practice." In *Ethics in Qualitative Research*, edited by Melanie Mauthner, Maxine Birch, Julie Jessop, and Tina Miller, 15–31. London: SAGE Publications Ltd, 2002.

Eizenberg, Efrat. *From the Ground Up: Community Gardens in New York City and the Politics of Spatial Transformation*. New York: Routledge, 2013.

Eliade, Mircea. *The Sacred and the Profane: The Nature of Religion*. New York: Houghton Mifflin Harcourt, 1959.

Eliasoph, Nina, and Paul Lichterman. "Culture in Interaction." *American Journal of Sociology* 108, no. 4 (2003): 735–794.

Emirbayer, Mustafa, and Ann Mische. "What Is Agency?" *American Journal of Sociology* 103, no. 4 (1998): 962–1023.

Fay, Brian. *Social Theory and Political Practice*. London: Routledge, 1975.

Fine, Gary Alan. "Small Groups and Culture Creation: The Idioculture of Little League Baseball Teams." *American Sociological Review* 44, no. 5 (1979): 733–745.

———. "The Storied Group: Social Movements as 'Bundles of Narratives.'" In *Stories of Change: Narrative and Social Movements*, edited by Joseph E. Davis, 229–245. Albany: State University of New York Press, 2002.

———. *Morel Tales: The Culture of Mushrooming*. Urbana: University of Illinois Press, 2003a.

———. "Towards a Peopled Ethnography Developing Theory from Group Life." *Ethnography* 4, no. 1 (2003b): 41–60.

Finney, Carolyn. *Black Faces, White Spaces: Reimagining the Relationship of African Americans to the Great Outdoors*. Chapel Hill: The University of North Carolina Press, 2014.

Firey, Walter. "Sentiment and Symbolism as Ecological Variables." *American Sociological Review* 10, no. 2 (1945): 140–148.

Foer, Jonathan Safran. *Eating Animals*. New York: Back Bay Books, 2010.

Frankenberg, Ruth. *White Women, Race Matters: The Social Construction of Whiteness*. Minneapolis: University of Minnesota Press, 1993.

Freudenburg, William R., Scott Frickel, and Robert Gramling. "Beyond the Nature/Society Divide: Learning to Think About a Mountain." *Sociological Forum* 10, no. 3 (1995): 361–392.

Fulkerson, Lee. *Forks Over Knives*. Virgil Films & Entertainment, 2011.

Futrell, Robert, Christie Batson, Barbara Brents, Andrea Dassopoulos, Chrissy Nicholas, Mark Salvaggio, and Candace Griffith. "Las Vegas Metropolitan Area Social Survey 2010 Highlights." 2010. http://digitalscholarship.unlv.edu/reports/2.

Gamson, William A. *The Strategy of Social Protest*. Homewood, IL: Dorsey Press, 1975.

Gardens For All. *Gardens for All: A Guide to a Greener, Happier Community*. Charlotte, VT: Gardens for All, 1973.

Gerlach, Luther P. "Movements of Revolutionary Change Some Structural Characteristics." *American Behavioral Scientist* 14, no. 6 (July 1, 1971): 812–836.

Gerlach, Luther P., and Virginia H. Hine. *People, Power, Change: Movements of Social Transformation*. New York: Bobbs-Merrill Company, 1970.

Gieryn, Thomas F. "Boundary-Work and the Demarcation of Science from Non-Science: Strains and Interests in Professional Ideologies of Scientists." *American Sociological Review* 48, no. 6 (1983): 781–795.

———. "A Space for Place in Sociology." *Annual Review of Sociology* 26 (2000): 463–496.

Gigliotti, Christina M., Shannon E. Jarrott, and Jeremy Yorgason. "Harvesting Health: Effects of Three Types of Horticultural Therapy Activities for Persons with Dementia." *Dementia* 3, no. 2 (2004): 161–180.

Glennie, Charlotte. "Gardens in the Growth Machine: Seattle's P-Patch Program and the Pursuit of Permanent Community Gardens." In *A Recipe for Gentrification: Food, Power, and Resistance in the City*, edited by Alison Hope Alkon, Yuki Kato, and Joshua Sbicca, 154–75. New York: NYU Press, 2020.

Glover, Troy D. "Social Capital in the Lived Experiences of Community Gardeners." *Leisure Sciences* 26, no. 2 (2004): 143–162.

Glover, Troy D., Kimberly J. Shinew, and Diana C. Parry. "Association, Sociability, and Civic Culture: The Democratic Effect of Community Gardening." *Leisure Sciences* 27, no. 1 (2005): 75–92.

Goffman, Erving. "The Nature of Deference and Demeanor." *American Anthropologist* 58, no. 3 (1956): 473–502.

———. *The Presentation of Self in Everyday Life*. New York: Anchor, 1959.

———. *Frame Analysis: An Essay on the Organization of Experience*. New York: Harper & Row Publishers, 1974.

Goffman, Erving, and Lyn H. Lofland. "On Fieldwork." *Journal of Contemporary Ethnography* 18, no. 2 (1989): 123.

Gorham, M. R., T. M. Waliczek, A. Snelgrove, and J. M. Zajicek. "The Impact of Community Gardens on Numbers of Property Crimes in Urban Houston." *HortTechnology* 19, no. 2 (2009): 291–296.

Gottdiener, Mark, Claudia Collins, and David R. Dickens. *Las Vegas: The Social Production of an All-American City*. Malden, MA: Wiley, 2000.

Gould, Kenneth A., and Tammy L. Lewis. *Green Gentrification: Urban Sustainability and the Struggle for Environmental Justice*. New York: Routledge, 2016.

Grasmuck, Sherri. *Protecting Home: Class, Race, and Masculinity in Boys' Baseball*. New Brunswick, NJ: Rutgers University Press, 2005.

Groenewegen, Peter P., Agnes E. Berg, Sjerp Vries, and Robert A. Verheij. "Vitamin G: Effects of Green Space on Health, Well-Being, and Social Safety." *BMC Public Health* 6 (2006): 149.

Guéguen, Nicolas, and Jordy Stefan. "'Green Altruism' Short Immersion in Natural Green Environments and Helping Behavior." *Environment and Behavior* 2014, 0013916514536576.

Guthman, Julie. "'If They Only Knew': Color Blindness and Universalism in California Alternative Food Institutions." *The Professional Geographer* 60, no. 3 (2008): 387–397.

Hall, John R., and Mary Jo Neitz. *Culture: Sociological Perspectives*. Englewood Cliffs, NJ: Prentice Hall, 1993.

Hanscom, Greg. "A Las Vegas Community Garden Beats the Odds." Grist, August 6, 2014. http://grist.org/cities/a-las-vegas-community-garden-beats-the-odds/.

Harmon, Laurie, and Laurel Harrington. "Building a Community Garden in Your Park: Opportunities for Health, Community, and Recreation." In *Grow Your Park Initiative*. Ashburn, VA: National Recreation and Park Association, 2010. https://www.nrpa.org/uploadedFiles/nrpa.org/Grants_and_Partners/Environmental_Conservation/Community-Garden-Handbook.pdf.

Helphand, Kenneth I. *Defiant Gardens: Making Gardens in Wartime*. San Antonio, TX: Trinity University Press, 2006.

Hemingway, John L. "Leisure, Social Capital, and Democratic Citizenship." *Journal of Leisure Research* 31, no. 2 (1999): 150–165.

Holstein, James A., and Jaber F. Gubrium. *The Active Interview*. Thousand Oaks, CA: SAGE Publications, 1995.

———. *The Self We Live By: Narrative Identity in a Postmodern World*. 1 edition. New York: Oxford University Press, 1999.

Hondagneu-Sotelo, Pierrette. "Cultivating Questions for a Sociology of Gardens." *Journal of Contemporary Ethnography* 39, no. 5 (2010): 498–516.

———. *Paradise Transplanted: Migration and the Making of California Gardens*. Berkeley: University of California Press, 2014.

Hou, Jeffrey, Julie M. Johnson, and Laura J. Lawson. *Greening Cities, Growing Communities*. Washington, DC: University of Washington Press, 2009a.

———. *Greening Cities, Growing Communities*. Washington, DC: University of Washington Press, 2009b.

Jackson, John B. *A Sense of Place, a Sense of Time*. New Haven: Yale University Press, 1994.

Jacobs, Jane. *The Death and Life of Great American Cities*. New York: Vintage, 1961.

Jamison, Michael S. "The Joys of Gardening: Collectivist and Bureaucratic Cultures in Conflict." *The Sociological Quarterly* 26, no. 4 (1985): 473–490.

Jasper, J. "A Strategic Approach to Collective Action: Looking for Agency in Social-Movement Choices." *Mobilization: An International Quarterly* 9, no. 1 (2004): 1–16.

Jasper, James M. *Getting Your Way: Strategic Dilemmas in the Real World*. Chicago: University of Chicago Press, 2006.

Jenkins, J. Craig. "Resource Mobilization Theory and the Study of Social Movements." *Annual Review of Sociology* 9, no. 1 (1983): 527–553.

Johnson, John M. "In-Depth Interviewing." In *Handbook of Interview Research: Context and Method*, edited by Jaber F. Gubrium and James A. Holstein, 103–119. Thousand Oaks, CA: SAGE Publications, 2002.

Johnstone, Barbara. *Stories, Community, and Place: Narratives from Middle America*. Bloomington, IN: Indiana University Press, 1990.

Kaplan, Rachel. "Some Psychological Benefits of Gardening." *Environment and Behavior* 5, no. 2 (1973): 145–162.

———. "The Nature of the View from Home Psychological Benefits." *Environment and Behavior* 33, no. 4 (2001): 507–542.

Kearney, Anne R. "Residential Development Patterns and Neighborhood Satisfaction Impacts of Density and Nearby Nature." *Environment and Behavior* 38, no. 1 (2006): 112–139.

Kingsley, Jonathan 'Yotti,' and Mardie Townsend. "'Dig In' to Social Capital: Community Gardens as Mechanisms for Growing Urban Social Connectedness." *Urban Policy and Research* 24, no. 4 (2006): 525–537.

Kingsley, Jonathan 'Yotti,' Mardie Townsend, and Claire Henderson-Wilson. "Cultivating Health and Wellbeing: Members' Perceptions of the Health Benefits of a Port Melbourne Community Garden." *Leisure Studies* 28, no. 2 (2009): 207–219.

Kobayashi, Audrey, and Linda Peake. "Racism Out of Place: Thoughts on Whiteness and an Antiracist Geography in the New Millennium." *Annals of the Association of American Geographers* 90, no. 2 (2000): 392.

Krieger, Alex. "Reinventing Public Space." *Architectural Record*, June 1, 1995.

Kudryavtsev, Alex, Richard C. Stedman, and Marianne E. Krasny. "Sense of Place in Environmental Education." *Environmental Education Research* 18, no. 2 (April 1, 2012): 229–250.

Kuletz, Valerie. *The Tainted Desert: Environmental Ruin in the American West*. New York: Routledge, 1998.

Kulin, Dan. "West LV Residents Rally for Grocery Store." Las Vegas Sun, July 8, 2005. http://lasvegassun.com/news/2005/jul/08/west-lv-residents-rally-for-grocery-store/#axzz2TsfGbkx6.

Kuo, Frances E., and William C. Sullivan. "Environment and Crime in the Inner City Does Vegetation Reduce Crime?" *Environment and Behavior* 33, no. 3 (2001): 343–367.

Kurtz, Hilda. "Differentiating Multiple Meanings of Garden and Community." *Urban Geography* 22, no. 7 (2001): 656–670.

Kusenbach, Margarethe. "Patterns of Neighboring: Practicing Community in the Parochial Realm." *Symbolic Interaction* 29, no. 3 (2006): 279–306.

Kweon, Byoung-Suk, Christopher D. Ellis, Pedro I. Leiva, and George O. Rogers. "Landscape Components, Land Use, and Neighborhood Satisfaction." *Environment and Planning B: Planning and Design* 37, no. 3 (2010): 500–517.

Lamont, Michèle, and Virág Molnár. "The Study of Boundaries in the Social Sciences." *Annual Review of Sociology* 28 (2002): 167–195.

Lang, John T. "Sound and the City." *Food, Culture & Society* 17, no. 4 (December 1, 2014): 571–589.

Lawson, Laura J. *City Bountiful: A Century of Community Gardening in America*. Berkeley: University of California Press, 2005.

Lefebvre, Henri. *The Urban Revolution*. Minneapolis: University of Minnesota Press, 2003.

Linthicum, Kate. "Plotting a Fee Increase at Community Gardens." *Los Angeles Times*, November 21, 2010. http://articles.latimes.com/2010/nov/21/local/la-me-community-garden-20101121.

Littlejohn, David, ed. *The Real Las Vegas: Life Beyond the Strip*. Oxford ; New York: Oxford University Press, 1999.

Lofland, John, David Snow, Leon Anderson, and Lyn H. Lofland. *Analyzing Social Settings: A Guide to Qualitative Observation and Analysis*. 4 edition. Belmont, CA: Wadsworth Publishing, 2005.

Lofland, Lyn H. *The Public Realm: Exploring the City's Quintessential Social Territory*. Hawthorne, NY: Aldine de Gruyter, 1998.

———. "Community and Urban Life." In *Handbook of Symbolic Interactionism*, edited by Larry T. Reynolds and Nancy J. Herman-Kinney, 937–974. Walnut Creek, CA: Rowman Altamira, 2003.

Logan, John R., and Harvey L. Molotch. *Urban Fortunes: The Political Economy of Place*. Berkeley: University of California Press, 1987.

Lohr, Virginia I., Caroline H. Pearson-Mims, and Georgia K. Goodwin. "Interior Plants May Improve Worker Productivity and Reduce Stress in a Windowless Environment." *Journal of Environmental Horticulture* 14 (1996): 97–100.

Lopez, Sandy. "Mobile Farmers Market Receives Grant to Spread to Henderson." *Las Vegas Review-Journal* (blog), February 15, 2018. https://www.reviewjournal.com/local/henderson/mobile-farmers-market-receives-grant-to-spread-to-henderson/.

Lowry, Christopher A., Jacob H. Hollis, Annick de Vries, B. Pan, Laura R. Brunet, Jon R. F. Hunt, Julian F. R. Paton, et al. "Identification of an Immune-Responsive Mesolimbocortical Serotonergic System: Potential Role in Regulation of Emotional Behavior." *Neuroscience* 146, no. 2 (2007): 756–772.

Lyon, Larry. *The Community in Urban Society*. Prospect Heights, IL: Waveland Press Inc, 1987.

Maines, David R. "Narrative's Moment and Sociology's Phenomena: Toward a Narrative Sociology." *The Sociological Quarterly* 34, no. 1 (1993): 17–38.

Maines, David R., and Jeffrey C. Bridger. "Narratives, Community and Land Use Decisions." *Social Science Journal* 29, no. 4 (1992): 363.

Mandela, Nelson. *Long Walk to Freedom: The Autobiography of Nelson Mandela.* Boston: Back Bay Books, 1994.

Marshall, Douglas A. "Temptation, Tradition, and Taboo: A Theory of Sacralization." *Sociological Theory* 28, no. 1 (2010): 64–90.

Martinez, Miranda J. *Power at the Roots: Gentrification, Community Gardens, and the Puerto Ricans of the Lower East Side.* Lanham, MD: Lexington Books, 2010.

Massey, Douglas S., and Nancy A. Denton. *American Apartheid: Segregation and the Making of the Underclass.* Cambridge, MA: Harvard University Press, 1993.

McKee, Robert, and Shannon M. Monnat. "Racialized Barriers and Social Action in West Las Vegas: The F Street Wall Controversy." *ASA Footnotes*, April 2011. https://www.asanet.org/sites/default/files/savvy/footnotes/apr11/vegas_0411.html.

McKelvey, Bill. "Community Gardening Toolkit: A Resource for Planning, Enhancing and Sustaining Your Community Gardening Project." University of Missouri Extension, 2009. http://extension.missouri.edu/explorepdf/miscpubs/mp0906.pdf.

Mears, Ashley. "Seeing Culture through the Eye of the Beholder: Four Methods in Pursuit of Taste." *Theory and Society* 43, no. 3–4 (2014): 291–309.

Melucci, Alberto. *Nomads of the Present: Social Movements and Individual Needs in Contemporary Society.* Philadelphia: Temple University Press, 1989.

———. *Challenging Codes: Collective Action in the Information Age.* Cambridge, MA: Cambridge University Press, 1996.

Milligan, Christine, Anthony Gatrell, and Amanda Bingley. "'Cultivating Health': Therapeutic Landscapes and Older People in Northern England." *Social Science & Medicine* 58, no. 9 (2004): 1781–1793.

Milligan, Melinda J. "Interactional Past and Potential: The Social Construction of Place Attachment." *Symbolic Interaction* 21, no. 1 (1998): 1–33.

Moehring, Eugene P. *Resort City in the Sunbelt: Las Vegas, 1930–2000.* Reno: University of Nevada Press, 2000.

Monti, Daniel J. *The American City: A Social and Cultural History.* Malden, MA: Blackwell Publishers, 1999.

Nettle, Claire. *Community Gardening as Social Action.* New York: Routledge, 2014.

Oldenburg, Ray. *The Great Good Place: Cafes, Coffee Shops, Bookstores, Bars, Hair Salons, and Other Hangouts at the Heart of a Community.* New York: Marlowe & Company, 1999.

Orum, Anthony M., and Zachary P. Neal. *Common Ground?: Readings and Reflections on Public Space.* New York: Routledge, 2009.

Packer, Adrienne. "Activists See Post-Integration Decay Continuing." Las Vegas Review-Journal, July 26, 2009. http://www.reviewjournal.com/news/activists-see-post-integration-decay-continuing.

Pantin, Lauren. "Chatting with Michael Pollan of The Omnivore's Dilemma." *Teen Vogue*, September 28, 2009. http://michaelpollan.com/interviews/chatting-with-michael-pollan-of-the-omnivores-dilemma/.

Park, Robert Ezra. "Human Ecology." *American Journal of Sociology* 42, no. 1 (1936): 1–15.

Park, Seong-Hyun, and Richard H. Mattson. "Therapeutic Influences of Plants in Hospital Rooms on Surgical Recovery." *HortScience* 44, no. 1 (2009): 102–105.

Parry, Diana C., Troy D. Glover, and Kimberly J. Shinew. "'Mary, Mary Quite Contrary, How Does Your Garden Grow?': Examining Gender Roles and Relations in Community Gardens." *Leisure Studies* 24, no. 2 (2005): 177–192.

Peterson, Kristen. "Westside Story: The Legacy and Heart of a Historic Neighborhood." *Las Vegas Weekly*, February 4, 2016. https://lasvegasweekly.com/news/2016/feb/04/westside-story-the-legacy-and-heart-of-a-historic/#/0.

Piven, Frances F., and Richard Cloward. *Poor People's Movements: Why They Succeed, How They Fail*. New York: Pantheon, 1977.

Pollan, Michael. *In Defense of Food*. New York: Penguin, 2008.

Polletta, Francesca. "Culture and Movements." *The Annals of the American Academy of Political and Social Science* 619, no. 1 (2008): 78–96.

Polletta, Francesca, and James M. Jasper. "Collective Identity and Social Movements." *Annual Review of Sociology* 27, no. 1 (2001): 283–305.

Pudup, Mary Beth. "It Takes a Garden: Cultivating Citizen-Subjects in Organized Garden Projects." *Geoforum* 39, no. 3 (2008): 1228–1240.

Putnam, Robert. *Bowling Alone: The Collapse and Revival of American Community*. New York: Simon & Schuster, 2000.

Riessman, Catherine K. *Narrative Analysis*. Newbury Park, CA: SAGE Publications, Inc, 1993.

Rothman, Hal. *Neon Metropolis: How Las Vegas Started the Twenty-First Century*. New York: Routledge, 2003.

Rowley, Rex J. *Everyday Las Vegas: Local Life in a Tourist Town*. Reno: University of Nevada Press, 2014.

Rupp, Leila, and Verta Taylor. *Drag Queens at the 801 Cabaret*. Chicago: University of Chicago Press, 2003.

Sanders, Scott R. *Staying Put: Making a Home in a Restless World*. Boston: Beacon Press, 1994.

Santlofer, Joy. "Asphalt Terroir." In *Gastropolis: Food and New York City*, edited by Annie Hauck-Lawson, Jonathan Deutsch, and Michael Lomonaco, 174–194. New York; Chichester: Columbia University Press, 2009.

Sbicca, Joshua. "The Urban Agriculture Fix: Navigating Development and Displacement in Denver." In *A Recipe for Gentrification: Food, Power, and Resistance in the City*, edited by Alison Hope Alkon, Yuki Kato, and Joshua Sbicca, 93–110. New York: NYU Press, 2020.

Scavone, Jason. "Overcoming Controversy on the Historic Westside." University of Nevada, Las Vegas News Center, July 1, 2016. http://www.unlv.edu/news/article/big-moves-salvage-redevelopment-efforts.

Schlosser, Eric. *Fast Food Nation: The Dark Side of the All-American Meal*. Boston: Mariner Books, 2012.

Schmelzkopf, Karen. "Urban Community Gardens as Contested Space." *Geographical Review* 85, no. 3 (1995): 364–381.

Schwalbe, Michael L., and Douglas Mason-Schrock. "Identity Work as Group Process." *Advances in Group Processes* 13 (1996): 113–147.

Shepard, Benjamin, and Greg Smithsimon. *The Beach Beneath the Streets*. Albany: State University of New York Press, 2011.

Shinew, Kimberly J., Troy D. Glover, and Diana C. Parry. "Leisure Spaces as Potential Sites for Interracial Interaction: Community Gardens in Urban Areas." *Journal of Leisure Research* 36, no. 3 (2004): 336–355.

Shubinski, Jennifer. "Councilman: Vons Closing 3 LV Stores." Las Vegas Sun, August 12, 2004. http://lasvegassun.com/news/2004/aug/12/councilman-vons-closing-3-lv-stores/.

Shupe, Ansom. "The Religious Institution." In *Handbook of Symbolic Interactionism*, edited by Larry T. Reynolds and Nancy J. Herman-Kinney, 625–636. New York: Rowman Altamira, 2003.

Smith, Christian. *Moral, Believing Animals*. Cambridge, MA: Oxford University Press, 2003.

Snow, David A. "Extending and Broadening Blumer's Conceptualization of Symbolic Interactionism." *Symbolic Interaction* 24, no. 3 (2001): 367–377.

Snow, David A., and Calvin Morrill. "Reflections on Anthropology's Ethnographic Crisis of Faith." *Contemporary Sociology* 22, no. 1 (1993): 8–11.

Snow, David A., and Doug McAdam. "Identity Work Processes in the Context of Social Movements: Clarifying the Identity/Movement Nexus." In *Self, Identity, and Social Movements*, edited by Sheldon Stryker, Timothy J. Owens, and Robert W. White, 41–67. Minneapolis: University of Minnesota Press, 2000.

Soga, Masashi, Kevin J. Gaston, and Yuichi Yamaura. "Gardening Is Beneficial for Health: A Meta-Analysis." *Preventive Medicine Reports* 5 (March 1, 2017): 92–99.

Staeheli, Lynn A. "Citizenship and the Problem of Community." *Political Geography* 27, no. 1 (2008): 5–21.

Stryker, Sheldon. "Identity Competition: Key to Differential Social Movement Participation?" In *Self, Identity, and Social Movements*, edited by Sheldon Stryker, Timothy J. Owens, and Robert W. White, 21–40. Minneapolis: University of Minnesota Press, 2000.

Swearingen, Scott. "Sacred Space in the City: Community Totems and Political Contests." In *Research in Community Sociology*, edited by Dan A. Chekki, 7: 141–169. Greenwich, CT: JAI Press, 1997.

Tavares, Stephanie. "West Las Vegas Group Converts Vacant Lot of Land to Community Garden." Las Vegas Sun, March 22, 2010. http://lasvegassun.com/news/2010/mar/22/community-garden-sprouts/.

Taylor, Verta. "Mobilizing for Change in a Social Movement Society." *Contemporary Sociology* 29, no. 1 (2000): 219–230.

Taylor, Verta, and Nancy E. Whittier. "Collective Identity in Social Movement Communities: Lesbian Feminist Mobilization." In *Frontiers in Social Movement Theory*, edited by Aldon D. Morris and Carol M. Mueller, 104–129. New Haven, CT: Yale University Press, 1992.

Teig, Ellen, Joy Amulya, Lisa Bardwell, Michael Buchenau, Julie A. Marshall, and Jill S. Litt. "Collective Efficacy in Denver, Colorado: Strengthening Neighborhoods and Health Through Community Gardens." *Health & Place* 15, no. 4 (2009): 1115–1122.

Three Square. "2018 Food Insecurity Map." Las Vegas, NV, 2018. http://www.threesquare.org/learn-about-hunger/research.

Totten, Kristy. "Garden Classrooms Project Sows Environmental Seeds." *Las Vegas Review-Journal*, July 12, 2013. http://www.reviewjournal.com/news/education/garden-classrooms-project-sows-environmental-seeds.

United States Department of Agriculture Economic Research Service. "Food Security in the U.S." 2018. https://www.ers.usda.gov/topics/food-nutrition-assistance/food-security-in-the-us/key-statistics-graphics.aspx#foodsecure.

Van Den Berg, Agnes E., Terry Hartig, and Henk Staats. "Preference for Nature in Urbanized Societies: Stress, Restoration, and the Pursuit of Sustainability." *Journal of Social Issues* 63, no. 1 (2007): 79–96.

Voicu, Ioan, and Vicki Been. "The Effect of Community Gardens on Neighboring Property Values." *Real Estate Economics* 36, no. 2 (2008): 241–283.

Wacquant, Loïc. *Body & Soul: Notebooks of an Apprentice Boxer*. Oxford; New York: Oxford University Press, 2006.

———. "Habitus as Topic and Tool: Reflections on Becoming a Prizefighter." *Qualitative Research in Psychology* 8, no. 1 (2011): 81–92.

Waliczekz, Tina M., Richard H. Mattson, and Jayne M. Zajicek. "Benefits of Community Gardening on Quality-of-Life Issues." *Journal of Environmental Horticulture* 14, no. 4 (1996): 194–198.

Weinstein, Netta, Andrew K. Przybylski, and Richard M. Ryan. "Can Nature Make Us More Caring? Effects of Immersion in Nature on Intrinsic Aspirations and Generosity." *Personality and Social Psychology Bulletin* 35, no. 10 (2009): 1315–1329.

Wellman, Barry, and Barry Leighton. "Networks, Neighborhoods, and Communities Approaches to the Study of the Community Question." *Urban Affairs Review* 14, no. 3 (1979): 363–390.

White, Mathew P., Ian Alcock, Benedict W. Wheeler, and Michael H. Depledge. "Would You Be Happier Living in a Greener Urban Area? A Fixed-Effects Analysis of Panel Data." *Psychological Science* 24, no. 6 (June 1, 2013): 920–928.

Williams, A. Park, Edward R. Cook, Jason E. Smerdon, Benjamin I. Cook, John T. Abatzoglou, Kasey Bolles, Seung H. Baek, Andrew M. Badger, and Ben Livneh. "Large Contribution from Anthropogenic Warming to an Emerging North American Megadrought." *Science* 368, no. 6488 (April 17, 2020): 314–318.

Wilson, Edward O. *Biophilia*. Cambridge, MA: Harvard University Press, 1984.

Wilson-Doenges, Georjeanna. "An Exploration of Sense of Community and Fear of Crime in Gated Communities." *Environment and Behavior* 32, no. 5 (2000): 597–611.

Winne, Mark. *Closing the Food Gap: Resetting the Table in the Land of Plenty*. Boston: Beacon Press, 2008.

Witcher, T. R. "Hello? Can We Get a Store Here?" *Las Vegas Weekly*, June 30, 2005. http://lasvegasweekly.com/news/archive/2005/jun/30/hello-can-we-get-a-store-here/.

Zelner, Sarah. "The Perpetuation of Neighborhood Reputation: An Interactionist Approach." *Symbolic Interaction* 38, no. 4 (2015): 575–593.

Zorbaugh, Harvey. "The Natural Areas of the City." In *The Urban Community: Selected Papers from the Proceedings of the American Sociological Society, 1925*, edited by Ernest W. Burgess, 219–229. Chicago: University of Chicago Press, 1926.

Zukin, Sharon. *Landscapes of Power: From Detroit to Disney World*. Berkeley: University of California Press, 1991.

———. *Naked City: The Death and Life of Authentic Urban Places*. New York: Oxford University Press, 2010.

Index

Page references for figures are italicized

ACGA. *See* American Community Gardening Association
adopt-a-plot, 36, 39–40, 70, *73*, 108, 114, 135; adopters, 41–42, 54, 58, 83–84, 87, 102, 108–9, 116, 128, 135, 139, 143, 150, 160–62, 166–67, *168*; cost of, 87, 134–36, 139, 143, 154
advisory board, 55–57, 59, 101–5, 110–15, 125–26, 131, 150, 152, 160–61, 164–66, *168*. *See also* working board
agency, 9, 16, 51, 69, 96, 98, 101, 103–4, 152, 154–55
alternative food movement, 12–15, 28, 32, 76–77, 120, 122, 127, 138–39, 151, 154–55
Amazon, 70, 143, 156n1
American Community Gardening Association, 65, 70, 75, 84
Aztec, 35, 79, *80–81*

Barlow, Ricki, 25, 27, 32–33, 76, 144
Basic Magnesium, 24, 30–31
Becker, Howard, 17, 67–68, 72, 74, 88–89, 91n9, 151, 169
belonging, 3–4, 55, 102–3, 119, 122–24, 133, 154
biophilia, 10
Black community, 29–33, 60, 120, 125, 127–32, 138–39, 170
Blee, Kathleen, 7, 9, 165
blind spots, 18, 60–61
boundaries: boundary work, 68, 120–23, 126–27, 129, 154; physical, 29, 31–32, 144; symbolic, 120, 122, 126, 129
Brooks, Rosalind "Roz," 37–43, 45n1, 49, 53–57, 59–60, 72–79, 83–88, 90, 97, 99–115, 120, 123–24, 126, 128–30, 132, 134, 136–38, 143, 145–56, 159, 163, 166, *168*, 169–70
bureaucracy, 99, 102–3, 106, 110

Caesars World, 25
caliche clay, 16, 162
church, 38, 40–41, 56, 87
Clair, Robin, 52
collective identity, 9, 60–61, 95–99, 102–3, 107, 115–16, 152–53
Colorado River, 24
community, 4, 6–7, 10–11, 15, 17–18, 21n46, 26, 28, 29, 31–33, 37, 39, 43–45, 47n41, 49, 51, 57, 60–61, 65, 69–73, 76–78, 80, 83–6, 88–90, 96–

97, 100–101, 111, 113, 119–33, 135, 137–39, 144–46, 148–52, 154–56, 159, 167, *168*; community lost, 3; community narrative, 55; community of believers, 56; community outreach, 111, 139, 148, 154, *168*; community places, 119, 122–23, 127, 129, 133; defining, 2–3, 5, 121; interpretive community, 55; practice of, 3, 45, 121
Community Development Corporation (CDC), 144
Community Development Programs Center, 38
community garden, 1–2, 5, 6, 9, 11–13, 15–17, 23, 25–27, 33, 36, 38–40, 42, 44, 46–7n41, 49–50, 53, 57–58, 65–80, 83–84, 87–89, 91n16, 91–92n18, 96–97, 100–101, 115, 119–23, 125, 127, 130–31, 134, 136, 138–39, 141n2, 145–49, 151, 154–56, 160, 167, 171
community service, 41–42, 108, *168*
community supported agriculture (CSA), 28, 146
compost, 36, 70, 71, 74, *75*, 92n25, 108, 134, 153, 162, *168*
cultural conventions, 5, 8, 17, 65–70, 72–74, 86–90, 91–92n18, 115, 151–52; linguistic and communicative conventions, 71, 74, 77, 79, 89, 151; material and spatial conventions, 70, 73–74, 78; organizational and administrative conventions, 71–72, 84, 86, 89, 151–52; overreliance, 17, 66, 72–79, 89–90, 152; selective reliance, 66, 72, 79–86, 89–90, 152
cultural production, 17–18, 65–68, 71–72, 79, 86–87, 89–90, 91n9, 91–2n18, 96, 151–52

Davis, Kingsley, 2
Davis, Mike, 6
day laborers, 5, 15–16, 39, 41–43, 83–84, 86, 90, 114–15, 152
Dignity Health, 143
displacement, 144–47
divine intervention, 53–56, 60–61
division of labor, 49, 72, 96, 99, 101–3, 115
Downtown Design Center, 144
Downtown Project, 144
drought. *See* water
Durkheim, Émile, 52, 58

Electronic Benefit Transfer (EBT), 143
Emirbayer, Mustafa, 9
exclusion, 119–23, 126, 133–34, 138–39; implicit exclusivity, 122, 133–37, 139, 154

farmers market, 13, 28, *35*, 36, 95, 114, 120, 136, 143, 146, 150, 153–55
Fine, Gary Alan, 51, 96
food desert, 17, 76, 154
food insecurity, 15, 17, 33, *34*, 76, 78, 132, 141n34, 143, 154
food justice, 12–14, 28, 76, 120, 139, 146, 151
food movement. *See* alternative food movement
Fremont Street, 34; East Fremont Street, 156n8; Fremont Street Experience, 3, 6
Fromm, Erich, 10

garden classes, 28, 35, 101, 112, 114, 143
Garden Farms, 28
Gardens for All, 70
gentrification, 145–46; green gentrification, 145–46. *See also* displacement
Gieryn, Thomas, 122
Goffman, Erving, 69, 80, 83, 86–87, 90, 164, 171
grassroots, 9, 12, 25, 50, 61–62, 69, 103, 106, 118n26, 146, 149, 152, 160, 171
Great Basin Permaculture Guild, 36

Green Our Planet, 28
Growing with Community Gardening, 70
Guthman, Julie, 14, 122

A Handbook of Community Gardening, 70
Harrah's, 25
Hawkins, Frank, 38
Hilton, 25
Historic West Las Vegas, 17–18, 28–34, 37–39, 42, 76–79, 88–89, 111, 113, 121, 125–26, 128, 130–32, 135, 137–39, 141n34, 144–47, 151, 153–54, 157n10, 167
Hondagneu-Sotelo, Pierrette, 15
Hoover Dam, 24
horizons of possibility, 51, 99, 111
Horsford, Steven, 32
HUNDRED Plan, 144–47, 156n3
hydrofetishism, 6

inclusion, 18, 61, 119–20, 122, 126, 129–31, 138–40, 154; explicit inclusivity, 122–26, 139, 154
irrigation, 36–37, 44, 70, 108–9, 134, *168*

Jasper, James, 97–98, 120, 123, 126
Jim Crow, 17, 28–30, 32

Krieger, Alex, 119
Kusenbach, Margarethe, 3

Lake Mead, 6
Las Vegas Culinary Academy, 144
Las Vegas Metropolitan Area Social Survey, 4
Latino community, 16, 33, 35, 60, 79, *81*–82, 125, 131, 139
Lil' Roots Garden Club, 114, 143
Lofland, Lyn, 3

Mandela, Nelson, 12
McWilliams Townsite, 30–31
MGM, 25
Mirage, 25
Mische, Ann, 9
mobilization, 23, 49, 71, 90, 96, 98, 147, 150. *See also* recruitment

narratives, 5–6, 14, 50–51, 55, 167; compatibility, 57–60; place narrative, 8, 50–57, 60–61, 79, 155; plot, 51, 55; spiritual, 17, 53–57, 60–61
National Association for the Advancement of Colored People, 38
Nellis Air Force Base, 24
Nevada Test Site, 6

organization dilemma. *See* strategic dilemma
organizational structure, 71, 90, 98–99, 101, 115–16, 150, 153, 164, 167, *168*; formal, 99–100, 103–5, 110, 112, 115, 152. *See also* grassroots; holocratic (organizational structure), 99, 118n27; informal, 99, 106–7, 115, 152–53; polycephalous (organizational structure), 99. *See also* bureaucracy

Patrick, Deval, 143
Pearson Community Center, 111
performativity, 66, 69, 71, 74, 79–80, 83, 86–87, 90, 151
permaculture, 36, 74, 84, 166, *168*
place, 4, 7–8, 16–18, 44, 45n2, 50–52, 55–56, 61–62, 65–70, 89, 90, 96, 116, 119–29, 133, 138, 144, 146–47, 151, 154–55, 160, 163, 165–66, 169–71; defining, 7; differential place, 146–47; performance of, 66, 69, 74, 79, 90, 151; place attachment, 4; place narrative. *See* narrative; placemaking, 65; sacred place. *See* sacred; space versus place, 121, 140n14; third place, 43
plot owners. *See* adopt-a-plot
plot renters. *See* adopt-a-plot

Pollan, Michael, 2, 14
power distribution, 72, 85, 90, 96, 99, 103, 140, 152–53, 165; centralized, 100, 102, 105–7, 113, 115, 152; decentralized, 105, 110, 152
prayer, 49, 53–54, 56
public, 43, 45n1–2, 71, 119–23, 125, 127, 133, 139, 170; versus community, 119–23, 133; defining, 119; public realm, 3

raised beds, 16, 36, 39, 41, 49, 70–71, *73*, 73–74, *75*, 86–87, 134, 136, 143, 161
recruitment, 39–40, 61, 66, 70, 87, 96, 100–101, 103, 106, 110–11, 130–31, 134, 154, 161. *See also* mobilization
Roz. *See* Brooks, Rosalind "Roz"

sacred, 52–53, 56–60, 63n26; and profane, 52, 58; sacralization, 52–54, 58–59; sacred narrative. *See* narrative; sacred place, 17, 49–50, 52–54, 56–60
Safran Foer, Johnathan, 14
Schlosser, Eric, 14
segregation, 5, 28–31, 77, 157n10
Shepard, Benjamin, 121–22, 133
shifting goals dilemma. *see* strategic dilemma
Sisolak, Steve, 26
Smith, Christian, 53
Smithsimon, Greg, 121–22, 133
Steyer, Tom, 143
strategic dilemmas and strategic choices, 9, 17, 50, 97–100, 103–5, 112–15, 120, 126, 149, 152–53; defining, 97–98; extension dilemma, 115, 120, 123, 126; organization dilemma, 98–99, 110, 115, 118n27, 152; shifting goals dilemma, 98, 110, 115, 152–53
the Strip (Las Vegas), 3, 18n5, 58, 157n10

Tonopah Community Garden, 5, 18n1, 112, 126, 159, 169
turning points

UCP. *See* urban culturalist perspective
University of Nevada, Las Vegas (UNLV)
urban culturalist perspective, 7–8, 17, 50, 154–55, 160
Urban Land Institute, 144
user-created, 84–86, 88–90, 151

Vegas SOUP, 113
Veggie Buck Truck, 114, 143
vermiculture, 36

water, 6, 10, 13, 15, 16, 24, 30–31, 35–36, 40, 43, 45n3, 69, 74, *75*, 83, 95, 108, 149, 155, *168*; drought, 6, 74; irrigation, 36–37, 44, 74, 108–9, 134; paying for, 26, 38, 87, 134, 145
Weekly, Lawrence, 32
working board, 87–88. *See also* advisory board

"you-pick," 13, 36, 95, 108, 136–37, 150, 153
Yucca Mountain, 6

Zappos, 99, 156n8
zoning, 26–27, 39

About the Author

Tyler S. Schafer is an assistant professor of Sociology at California State University, Stanislaus.

www.ingramcontent.com/pod-product-compliance
Lightning Source LLC
LaVergne TN
LVHW011814200926
848484LV00005B/1197

* 9 7 8 1 7 9 3 6 2 3 1 2 6 *